W9-CBM-676

DIFFUSE X-RAY REFLECTIONS FROM CRYSTALS

W. A. WOOSTER

DOVER PUBLICATIONS, INC.
Mineola, New York

Bibliographical Note

This Dover edition, first published in 1997, is an unabridged and unaltered republication of the work first published by The Oxford University Press, Oxford, in 1962.

Library of Congress Cataloging-in-Publication Data

Wooster, W. A. (William Alfred)
Diffuse X-ray reflections from crystals / W. A. Wooster.
p. cm.
Originally published: Oxford: Clarendon Press, 1962.
Includes bibliographical references and indexes.
ISBN 0-486-69641-3 (pbk.)
1. X-ray crystallography. 2. X-rays—Diffusion. I. Title.
QD945.W6 1997
548'.83—dc21 97–13551
CIP

Manufactured in the United States of America
Dover Publications, Inc., 31 East 2nd Street, Mineola, N.Y. 11501

TO
PAUL P. EWALD

PREFACE

THE SUBJECT of diffuse X-ray reflections from crystals has acquired the reputation of being difficult and mathematical. It has been one of the aims of the author in writing this book to follow a dictum of Lord Rutherford, namely, that every important idea must be capable of simple expression. The emphasis here has been put on the underlying physical ideas, though the author hopes that the analysis has also received adequate treatment. The reader whom the author has tried to help is the young research student.

The author has experienced difficulty in selecting material to be included in this book. Electron and neutron diffraction have, except for one reference, been excluded, although there are many important relations between these types of diffraction and those considered here. It was felt better to restrict the subject matter to one form of diffraction for which the experimental techniques are all practised within the same type of laboratory. A similar difficulty has arisen in connection with some theoretical treatments. The method developed by Born and his colleagues has been largely omitted because the phenomena of thermal diffuse scattering can be correlated in a different way by the simpler treatment which originated with Faxen and Waller. The subject of crystal dynamics has not been dealt with here except in its relationship to the phenomena of diffuse X-ray reflection. In a complete treatment of the thermal vibrations of atoms this would be a serious omission but it was felt better in this work to leave the student to refer to the classical works rather than to attempt any summary of them.

Not all so-called diffuse effects have been dealt with in this book. One notable omission is the type associated with the precipitation of very small but parallel orientated crystals of one phase within a matrix of another phase. These effects are not strictly diffuse by virtue either of thermal motion or of static atomic displacements and they were omitted for this reason.

The examples of actual problems have been chosen so as to illustrate the general principles and have been worked out in sufficient detail to permit a student to carry out the analysis of the experimental data for himself. Some tables of actual experimental results have been given for this reason. No attempt has been made to cover all known examples but it is hoped that a sufficiently representative choice of examples has been given.

The problem presented by the choice of symbols has been difficult. There is rather little in common between the systems of symbols used by different authors. As yet there is no international convention. In this work quantities related directly to the reciprocal lattice are denoted by an asterisk after the symbol. Some years ago the special terms 'relp', 'rel-vector', and 'rekha' were introduced to avoid the constant repetition of fairly long phrases. These terms have been found useful and are retained in this work.

The author is greatly indebted to his pupils and colleagues who since 1948 have worked with him in this field, namely, Professor G. N. Ramachandran, Dr Andrew Land, Mr G. L. Macdonald, Dr J. A. Hoerni, Dr E. Prince, Dr S. C. Prasad, and Dr E. Sándor. The technical assistance from Mr C. Chapman of the Cavendish Laboratory, Messrs N. A. Lanham, K. Rickson, and R. Lee of the Department of Mineralogy and Petrology, and Messrs J. A. L. Fasham, D. Day, and I. Perry of the Brooklyn Crystallographic Laboratory is gladly acknowledged. For help with the lettering of the diagrams the author wishes to thank his son, Mr G. A. Wooster.

Cambridge,
Jan. 1961

W.A.W.

CONTENTS

CHAPTER I: GENERAL SURVEY

CHAPTER II: THE DETERMINATION OF ELASTIC CONSTANTS OF CRYSTALS BY DIFFUSE X-RAY REFLECTIONS

I

GENERAL SURVEY

1:1 Historical review

THE GENERAL diffuse scattering of X-rays from solid materials has been known from the times of the earliest work on the interaction of X-rays with matter. Experimental crystallographers have generally regarded this diffuse scattering as an unwanted effect which produced a background darkening on their X-ray photographs or an equivalent effect in their diffractometers. During the last two decades it has been shown that part of this diffuse scattering can provide useful information about many physical properties of crystals. The scattering or reflection of X-rays, of the type known as Laue or Bragg reflections, which gives rise to spots in single-crystal X-ray photographs, is due to the interaction with the X-rays of atoms arranged on a regular periodic lattice. Here we shall be concerned either with lattices which in one way or another depart from this regular periodic character, or with relatively small groups of atoms having a certain pattern within the group. Such groups of atoms will generally be distributed in a more or less regular way over a lattice. Thus we shall exclude from this survey the scattering by gases, liquids, and truly amorphous solids; the range to be covered lies between such materials and the perfect crystal at absolute zero temperature.

Debye (1913*a*, *b*, *c*, 1914) developed a theory concerning the diminution of the intensity of X-ray reflections due to the thermal vibration of atoms. Though the main purpose of this work was concerned with normal X-ray reflections it had in fact a close connection with the diffusely reflected part of the radiation. This work was carried a stage further by Faxen (1923) and by Waller (1923, 1925, 1928), who took account of the anisotropic elastic character of crystals. It was shown that the thermal motion of the atoms at all temperatures above absolute zero could be regarded as due to the superposition of a great number of elastic waves of many frequencies and directions of travel. On this basis it was possible to derive the diminution of the intensity of the normal X-ray reflections. This work was followed up in connection with experimental studies on the variation with temperature of reflections from certain crystals of simple crystal structures (Brindley and Ridley, 1938) but the other aspect of the work, namely, that concerned with the diffusely scattered X-rays themselves, was not studied until Laval (1939) published his first work in this field. Zachariasen (1940), independently of Laval, developed substantially the same theory and tested it by measurements on rock salt (Siegel and Zachariasen, 1940). Hall (1942) carried out similar tests on potassium chloride and potassium bromide. There began at this time a controversy concerning the diffuse scattering of diamond which was to extend over a number of years. Raman and his co-workers developed a theory which had some features in common with the papers by Laval and Zachariasen: Raman and Nath (1940*a*, *b*), Raman and Nilakantan (1940*a*, *b*, *c*), Raman (1941*a*, *b*), Raman and Nilakantan (1941*a*, *b*), Pisharoty (1941), Pisharoty and Subrahmanian (1941), Raman (1942), Nath (1943), Krishnan and Ramachandran (1945), Raman (1948, 1955, 1958*a*, *b*, *c*). In one respect, however, the Indian work differed strongly from the rest. Whereas Faxen, Waller, and following them, Laval, Zachariasen, and Born and Sarginson (1941) had attributed the diffuse scattering to the effect of elastic waves which continually travel through the crystal, Raman supposed that the X-rays themselves generated the waves which gave rise to the diffuse scattering. The waves so generated were supposed to be directly related to the infra-red spectra of the crystals in which they were produced. Subsequent work has shown that in general the phenomena of diffuse scattering can be divided into two groups. In the first group the effects are explained by the thermally excited elastic waves, as assumed originally by Faxen and Waller, and in the second group the effects are explained by particular features of the static atomic arrangement in the crystal. Diamond shows both types of diffuse scattering (Lonsdale, 1942*a*), and this added to the difficulty of resolving the controversy. This particular matter has been reviewed by Born (1942*a*, *b*) and Lonsdale (1942*c*, 1945), and comprehensive reports on the experimental and theoretical aspects of the whole subject were given by Lonsdale (1943) and Born (1943) respectively. The theory of thermal diffuse scattering was further developed by Begbie and Born (1947) and by Begbie (1947). Studies on elastic spectra were carried out on aluminium by Olmer (1948), on alpha-iron by Curien (1952*a*, *b*, *c*, *d*), on beta-brass by Cole and Warren (1952), and on zinc by Joynson (1954). The elastic spectrum of copper was determined by Jacobsen (1955) and that of aluminium was re-determined by Walker (1956). Since the original work of Laval and Olmer a large number of authors have made determinations of elastic constants by studying thermal diffuse scattering and this is more fully described in Chapter II.

Parallel with the work on thermal diffuse scattering there was much research on structural diffuse scattering. This began in 1928, almost at the same time as the study of thermal diffuse scattering, and has developed along a number of lines. These are discussed in Chapter IV but here we may mention the main themes, namely, (*a*) layer structures such as those of the micas, (*b*) close-packed metals, including cobalt, (*c*) diamond, (*d*) age-hardening alloys, (*e*) various order-disorder problems in inorganic and metallic crystals. The study of solid solutions both theoretically and experimentally has been much developed. Flinn and Averbach (1951) investigated solid solutions of gold and nickel. Taylor (1951*a*, *b*) gave a theoretical treatment of disorder in binary alloys and other authors have studied the effect on diffuse scattering of various types of lattice defect: van Raijen (1944), Ekstein (1945), Guinier (1945*a*, *b*), Lonsdale (1945), Burgers and Hiok (1946), Deas (1952), Kakinoki and Komura (1952), Matsubra (1952), Paterson (1952), Walker (1952), Wilson (1952), and Krivoglaz (1957, 1958).

Lastly, diffuse reflection studies have been applied to molecular crystals. This is discussed in Chapter V and here we may simply note that this method promises to throw light on the shape and orientation of molecules and may also become a useful auxiliary method in the determination of crystal structures.

Thus the study of diffuse X-ray reflections began as a means of accounting for the dependence of the intensity of normal Bragg-Laue reflections on temperature. This led to the theory of the connection with the elastic properties of crystals and of the force constants between the constituent atoms. The phenomena of diffuse X-ray reflection have been divided into those concerned with atomic vibrations and those concerned with static atomic displacements. The researches on the second group have led to a deeper understanding of (*a*) age-hardening alloys in which actual or incipient precipitation occurs, and (*b*) the defects associated with solid solutions and dislocations. At the end of this period of rapid development has come the application of the method to the study of the shape and orientation of molecules in organic crystals.

1:2 **Theoretical basis of diffuse reflection studies**

1:2.1 *General relations*

A demonstration of the kind of changes produced in an X-ray diffraction pattern by various disturbances of the regularity of a crystal lattice can be given by optical means. The apparatus used is well known in connection with the determination of crystal structures and is often called an optical diffractometer. It consists, as shown in Fig. 1.1, of a lamp, *A*, pinhole, *B*, either one or

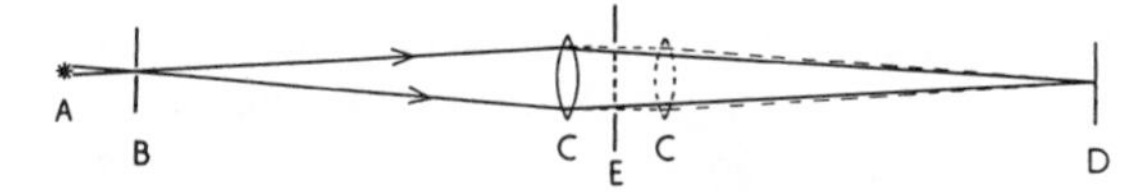

FIG. 1.1 Diagram representing the component parts of an optical diffractometer. *A*, light source; *B*, pinhole aperture; *C*, *C*, long focus lenses; *D*, screen or photographic film; *E*, mask with holes in an opaque screen.

two long-focus lenses, *C*, *C*, and a screen, *D*, or camera. The pinhole and screen are at equal distances from the lens (or lenses) and placed so that the pinhole is sharply focused on the screen. If a mask, *E*, having holes distributed in any way, is placed next to the lens (or between the two lenses), a diffraction pattern corresponding to the mask is obtained on the screen. If the distance from lens to screen is about two metres then a mask having a square lattice of holes

1 mm apart produces a square array of diffraction spectra also 1 mm apart. To obtain the sharpest diffraction pattern it is necessary to use monochromatic light.

To show the effect of various types of disturbances of the lattice, a number of masks were made by photographing the corresponding pattern of dots drawn on paper. These patterns are shown in Figs. 1.2*a*–1.7*a*. It will be seen that the same shape of molecule (naphthalene-like) in the same orientation has been used throughout. Figs. 1.2*b*–1.7*b* give the diffraction patterns produced

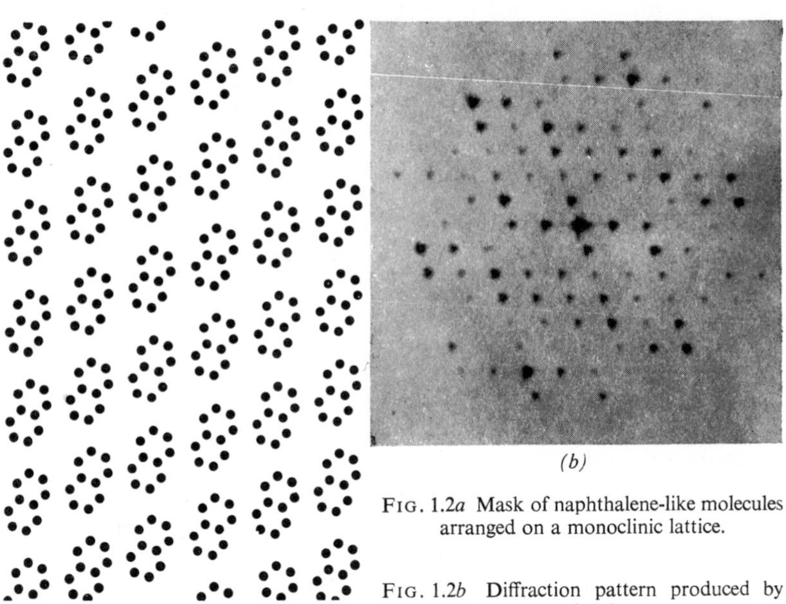

FIG. 1.2*a* Mask of naphthalene-like molecules arranged on a monoclinic lattice.

FIG. 1.2*b* Diffraction pattern produced by mask 1.2*a*.

by these masks. It will be seen that the diffraction spectra in all figures 1.2*b*–1.7*b* are arranged on a lattice reciprocal to that of the patterns 1.2*a*–1.7*a*. The various patterns are arranged so as to show the effects on the diffraction of different types of disturbance of the lattice.

Fig. 1.2*a* shows one-ninth part of the pattern which formed the mask corresponding to a perfect crystal and Fig. 1.2*b* the corresponding series of diffraction spots. In Fig. 1.3*a* is shown the pattern in which all the molecules have their axes parallel but their centres are arranged in a random manner with respect to one another. The resulting diffraction pattern is shown in Fig. 1.3*b* and this corresponds approximately to the effect which would be produced by a mask having one single molecule. The diffraction pattern is, as we shall see (Figs. 5.1*a*, *b*), the transform of the pattern of the molecule itself (Lipson and Taylor, 1958). In Fig. 1.4*a* is shown the regular lattice

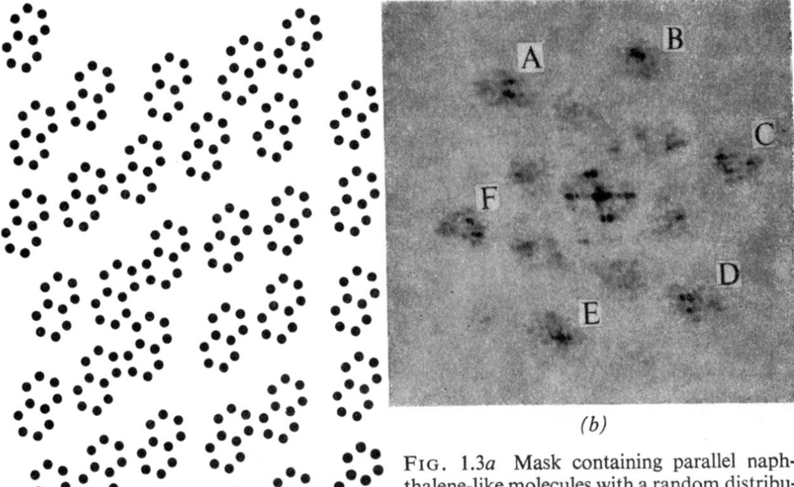

FIG. 1.3*a* Mask containing parallel naphthalene-like molecules with a random distribution of their centres.

FIG. 1.3*b* Diffraction pattern produced by mask 1.3*a*

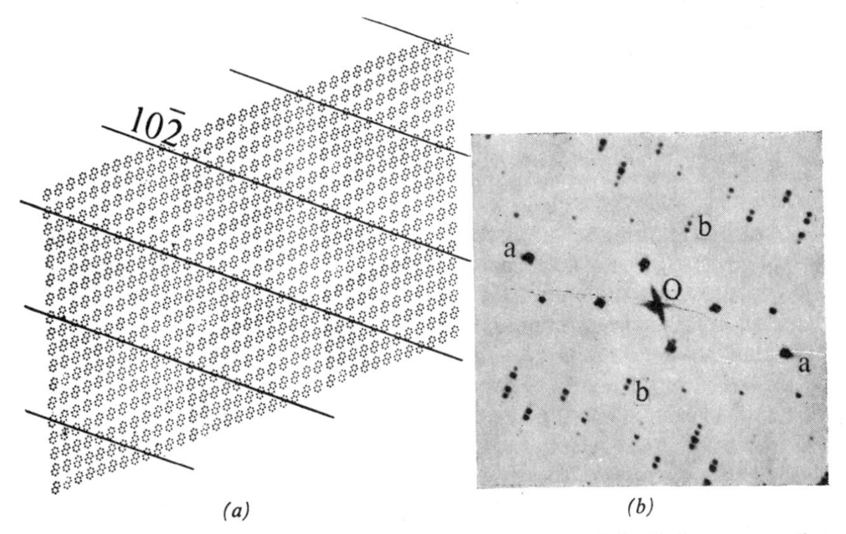

FIG. 1.4*a* Mask of naphthalene-like molecules having a periodic displacement so that rows of molecules in the same phase are parallel to ($10\bar{2}$).

FIG. 1.4*b* Diffraction pattern due to mask 1.4*a*, showing the satellites on either side of the main spectra.

disturbed by the passage through it of elastic longitudinal waves parallel to the crystallographic plane ($10\bar{2}$). It can be seen that most of the corresponding diffraction spots in Fig. 1.4*b* have satellites. Along the line *aa*, which is parallel to the wave fronts of the elastic waves marked with lines in Fig. 1.4*a*, there are no satellites. This corresponds to the fact that the longitudinal waves do not disturb the lattice in a direction perpendicular to the wave normal. The points *bb*, of indices $10\bar{2}$, $\bar{1}02$ respectively, have only one satellite on each side of the main diffraction maximum. The same is true for the maxima along a line through *b* parallel to the line *aa*. The line joining the two satellites on

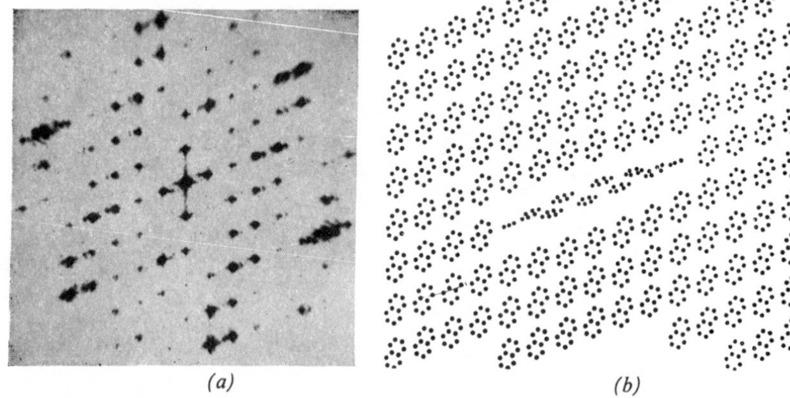

(*a*) (*b*)

FIG. 1.5*a* Mask illustrating a lattice with an inclusion arranged parallel to plane (001).

FIG. 1.5*b* Diffraction pattern due to mask 1.5*a*, showing streaks parallel to c^* passing through the main spectra. (The author is indebted to Dr C. A. Taylor for preparing this photograph.)

either side of the point *b* is normal to the wave front $10\bar{2}$ in Fig. 1.4*a*, and the separation of the satellite from its original diffraction maximum is one-twelfth of the distance of 0*b*. It will be seen from Fig. 1.4*a* that there are twelve parallel planes of corresponding layers of molecules between each pair of wave fronts. It is a universal rule that the distance of the satellite from its associated original maximum is inversely proportional to the wavelength of the wave motion disturbing the lattice. Along lines parallel to *aa* but further from 0 than *b*, there are two or more satellites on either side of the original maximum. This has its counterpart in the diffraction of X-rays. The line passing through the satellites is again perpendicular to *aa* and the separation is everywhere the same as for the point *b*.

In Fig. 1.5*a* a mask in which the effect of a platelet disturbing the lattice parallel to the given plane is shown. In the resulting diffraction pattern (Fig. 1.5*b*) each spot has a spike sticking out roughly in a direction normal to the length of the platelet. It should be noticed that every spot in Fig. 1.5*b* is similarly affected. In Chapter IV we shall encounter examples of such platelets and the corresponding X-ray pattern. In Fig. 1.6*a* (which is only one-ninth of the whole pattern) all the molecules have their centres arranged on the regular lattice and their axes are all parallel to one another, but the molecules are supposed to be vibrating along their lengths. The length of any given molecule has been chosen in a random manner. In other words, the pattern of Fig. 1.6*a* is meant to represent the thermal vibration of these molecules

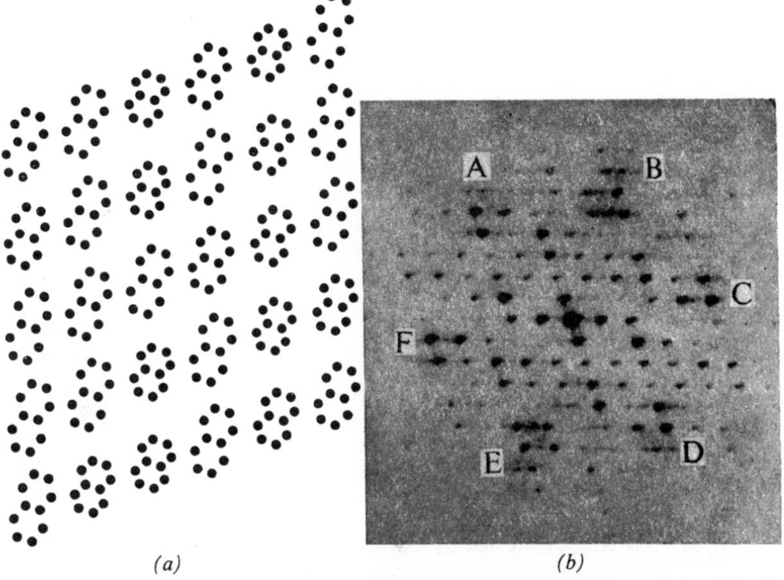

FIG. 1.6*a* Mask illustrating naphthalene-like molecules having their centres placed in the undisturbed lattice but their lengths varying at random.

FIG. 1.6*b* Diffraction pattern due to mask 1.6*a*, showing the development of features corresponding to Fig. 1.3*a*.

along their lengths, each one being supposed independent of its neighbours. The resulting diffraction pattern is shown in Fig. 1.6*b* and it will be seen that although the points of the pattern are in general sharp there are regions, such as those marked *A*, *B*, *C*, *D*, *E*, *F*, which correspond to those regions similarly marked in Fig. 1.3*b*. In other words, superimposed upon the regular diffraction pattern of the lattice is the transform of the individual molecule. In Fig. 1.7*a* is shown a part of the mask where the centres of all the molecules are arranged on a regular lattice, where their shapes are all identical, but their axes have been made to vibrate in a random manner up to 20° on either side of the normal direction corresponding to Fig. 1.2*a*. The resulting diffraction pattern is shown in Fig. 1.7*b*. Superimposed on the normal diffraction pattern (Fig. 1.2*b*) is a diffuse reflection which is roughly what would be obtained by rotating Fig. 1.3*b* about its centre.

The diffraction patterns (Figs. 1.2*b*–1.7*b*) illustrate most of the problems dealt with in diffuse reflection studies. The serious omission is that of order-disorder problems, but these will be discussed in Chapter IV.

In Fig. 1.4*a* there is only one wave perturbing the lattice but if we suppose that there are a number of waves having a common wave normal but differing wavelengths, then to each diffraction spot there will be a number of satellites

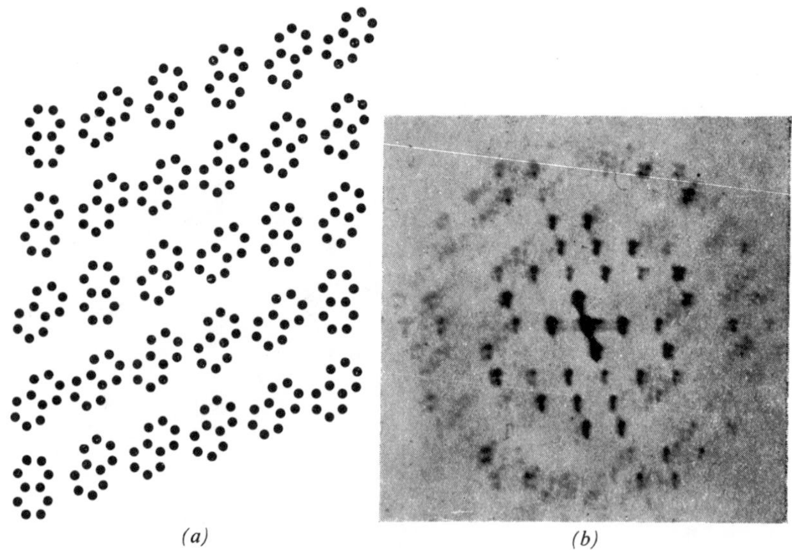

(*a*) (*b*)

FIG. 1.7*a* Mask representing molecules having their centres placed on an ideal lattice but their lengths distributed at random about the mean direction.

FIG. 1.7*b* Diffraction pattern due to mask of 1.7*a*.

arranged along the same line through the diffraction spot at distances from it inversely proportional to the corresponding wavelengths of the waves producing them. Owing to the thermal agitation that atoms experience at ordinary temperatures there are an enormous number of elastic waves travelling through the crystal backwards and forwards and each one of these produces its own satellite around each diffraction spot. In real crystals the number of these satellites is so large that they merge into a continuous cloud and round each diffraction spot such as those shown in Fig. 1.2*b* we obtain a region of diffuse reflection which may be associated with these elastic waves. The connection between elastic waves travelling through any medium and the elastic properties of the medium was first worked out a long time ago and is subject to the laws of elasticity for the medium concerned. Elastic properties of crystals are usually anisotropic and the observation of the diffuse reflection surrounding any one of the main reflections can be made to yield some or all of the elastic constants. This is more fully dealt with in Chapter II.

When all the elastic properties have been thoroughly studied it is possible to determine the elastic spectrum for the crystal, that is to say, the distribution of the wavelengths and intensities of the waves which are continually travelling to and fro in the crystals at different temperatures. These elastic spectra have a fundamental importance and are involved in specific heat and other studies of the physical properties of crystals. These will be dealt with more fully in Chapter III.

When the atoms or molecules of a lattice are permanently displaced in a random manner from the positions that they would occupy in a perfect lattice we are confronted by the problem of 'order-and-disorder'. The effect of such random displacements on the diffuse intensity is to cause a distribution of diffuse scattering in all directions and especially in directions far removed from the main Bragg reflections. Whereas the elastic properties can only be studied very close to the Bragg reflection, the regularity of arrangement of the atoms or molecules can best be studied at points far removed from any of the main reflections. This is more fully dealt with in Chapter IV.

When impurities are not randomly distributed as isolated atoms or molecules but are segregated into plates or rods then we have the case represented by Fig. 1.5*a*. The consequent distortion of the lattice is, of course, a static one but it can be resolved, at any rate in most cases, into the equivalent of a series of plane waves all parallel to the plane of the platelet or to the length of the rod. The diffraction pattern due to such plane waves consists of spikes corresponding to a succession of satellites distributed along the direction normal to the wave front. The analysis of the intensity distribution within such satellites can give information about the nature of the precipitated platelets. Such problems are common in age-hardening alloys and in a number of mineral crystals in which two phases separate from one another during the course of growth. These problems are more fully dealt with in Chapter IV.

Thermal agitation in crystals results in the translation of molecules, due to the passage of elastic waves through the lattice, and also in the vibration of molecules—vibrations which change the molecular length or rotate the axes of the molecules relative to their normal positions. Fig. 1.6*a* shows the type of movement associated with vibration and Fig. 1.7*a* shows that associated with oscillation. The effects of such vibration or oscillation are felt in the diffuse reflection. The orientation and some other geometrical features of the molecular Fourier transform can be found from the experimental observations. From this the orientation and shape of the molecule itself can be derived.

1:2.2 *Analytical statement*

The optical diffractogram of Fig. 1.4*b* showed that waves of wavelength long compared with the cell dimensions of the lattice, when imposed upon the lattice, gave rise to satellites in the resulting diffraction pattern. It is our object at this point to see how this phenomenon arises. We shall first consider two elementary problems, namely, (*a*) the effect on the scattering of X-rays of a medium which has a purely sinusoidal distribution of density, and (*b*) the effect of imposing a second sinusoidal distribution on the first, the wavelength of the second distribution being long compared with that of the first. These elementary cases illustrate the essence of the problem presented by a lattice disturbed by waves, because the electron density within the unit cell can be resolved into a series of sinusoidal distributions corresponding to case (*a*). In the undisturbed crystal each X-ray reflection of the normal Bragg kind may be regarded as an example of case (*a*). In the lattice perturbed by the elastic waves the normal X-ray reflection is flanked by satellites, each of which is associated with one elastic wave. This is the result which is shown in its simplest form by a consideration of case (*b*).

1:2.2.1 *Diffraction by parallel planes having a sinusoidal distribution of density*

X-rays are supposed to fall on parallel planes of scattering matter along the line *PO* (Fig. 1.8). Reflection occurs at the same angle as the angle of incidence, θ, from all the successive layers of scattering matter. The top layer is denoted by *AB* and any lower layer by *CD*. The perpendicular distance between the planes *AB*, *CD* is denoted x. The path difference between the rays

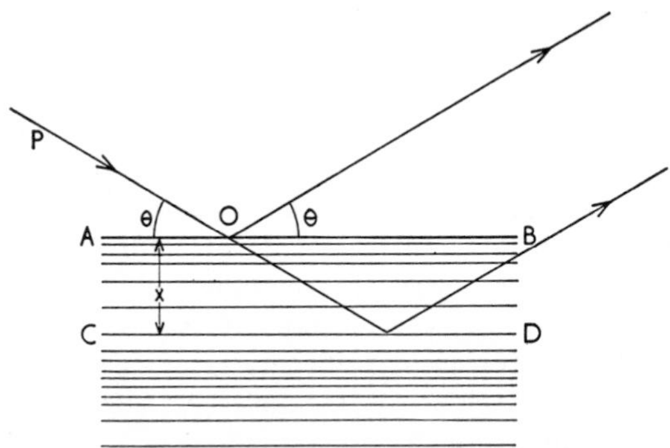

FIG. 1.8 Diagram of rays reflected from parallel strata having sinusoidal displacements from their original equally-spaced positions.

reflected from planes *AB* and *CD* is $2x \sin \theta$, and the corresponding phase difference is $2\pi . 2x \sin \theta/\lambda$, where λ is the wavelength of the X-rays. We shall suppose that the density of scattering matter at the level *CD* is ρ_x. Then the amplitude of the wave scattered from a layer of thickness dx at level *CD* will be proportional to

$$\rho_x dx.$$

If ρ_x varies sinusoidally let us set

$$\rho_x = \rho + \rho_1 \cos 2\pi x/L,$$

where ρ is the average electron density, ρ_1 is the amplitude of the sinusoidal variation of electron density, and L is the wavelength of the density-waves.

Thus the product of the amplitude and the cosine of the phase angle corresponding to the scattering from the thin layer at level CD is

$$(\rho+\rho_1 \cos 2\pi x/L)dx.\cos 2\pi.2x \sin \theta/\lambda.$$

When this expression is integrated from $x = 0$ to $x = L$ we obtain the scattering for a thickness L as

$$\tfrac{1}{2}\rho_1 L.$$

This result can be appreciated simply by considering the product of the ordinates of two sinusoidal curves of different wavelengths. The curve, corresponding to the product at each point along the x-axis of the two cosine waves, is above and below the x-axis for equal lengths along that axis. Taking areas between the product curve and the x-axis to be positive when above the x-axis and negative when below it, we see that the algebraic sum of the areas between the product curve and the x-axis is zero over each length along the x-axis which corresponds to a repeat unit of the product curve. Only when the wavelengths of the two cosine waves are equal is the product of the ordinates of the same sign at all points along the x-axis. If the common wavelength is L and the amplitude unity, then the area under the product curve in a length L is $\frac{1}{2}L$.

To make the wavelengths of the two waves equal in the case discussed here we must impose the condition

$$x/L = 2x \sin \theta/\lambda$$

or,

$$L = \lambda/2 \sin \theta. \tag{1.1}$$

Thus a sinusoidal distribution of scattering matter of a given L-value can only scatter X-rays at a particular value of θ given by equation (1.1). It should be noted that, in contrast with the scattering from the parallel planes of a lattice, there is here no possibility of second or higher order reflections. One and only one reflection is possible and the amplitude is proportional to ρ_1 and the thickness irradiated.

1:2.2.2 *Diffraction by parallel planes having a sinusoidal distribution of density which is perturbed by a longitudinal wave of long wavelength*

We shall assume that the distribution of the scattering matter is the same as that assumed in section 1:2.2.1 except that a second wave, also parallel to the reflecting surface, AB, but of longer wavelength is superimposed on the original distribution. The wavelength of the second wave is taken as Λ and its amplitude as ρ'. In Fig. 1.9 the density variation with x in the absence of the perturbing wave is shown as a full line. The effects of the perturbing wave are two-fold: for a given value of x there is a displacement of the density distribution to a position along the x-axis shown by the dotted line (Fig. 1.9)

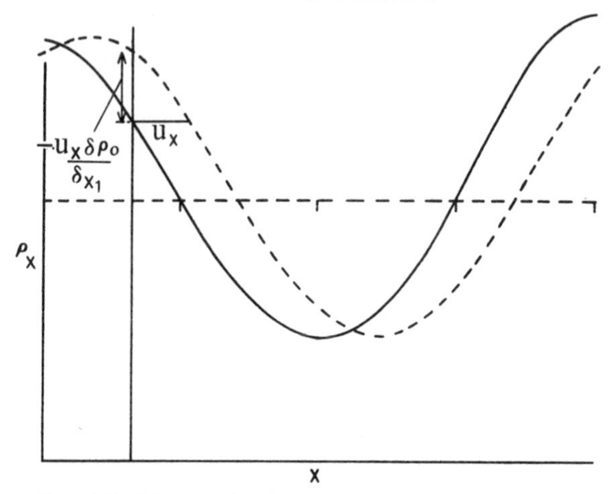

FIG. 1.9 Diagram showing the variation of electron density along a wave normal due to the passage of an elastic wave.

and also an expansion or contraction of the scattering material. The first effect increases the density by an amount $-u_x\partial\rho_0/\partial x$ where u_x is the displacement of the L-curve at the depth x and ρ_0 is the original density of a point x. The second effect changes the density by a factor $(1-\partial u_x/\partial x)$. Thus the density at a depth x is given by

$$\rho_x = (\rho_0 - u_x\,\partial\rho_0/\partial x)(1-\partial u_x/\partial x),$$

Now

$$\rho_0 = \rho+\rho_1 \cos 2\pi x/L,$$
$$\partial\rho_0/\partial x = -\rho_1 . \frac{2\pi}{L} . \sin 2\pi\frac{x}{L},$$
$$u_x = \rho' . \cos 2\pi x/\Lambda.$$

Hence

$$\partial u_x/\partial x = -\rho'\frac{2\pi}{\Lambda} . \sin 2\pi\frac{x}{\Lambda}.$$

Hence

$$\rho x = \left(\rho+\rho_1 \cos 2\pi\frac{x}{L} + \rho' \cos 2\pi\frac{x}{\Lambda} . \rho_1 . \frac{2\pi}{L} . \sin 2\pi\frac{x}{L}\right)\left(1+\rho'\frac{2\pi}{\Lambda} . \sin 2\pi\frac{x}{\Lambda}\right)$$

and, neglecting the product $(\rho')^2\rho_1$, which has a small value, we obtain

$$\rho_x = \rho+\rho_1 \cos 2\pi\frac{x}{L} + \rho'\rho_1\frac{2\pi}{L} . \cos 2\pi\frac{x}{\Lambda} . \sin 2\pi\frac{x}{L} +$$
$$+\rho\rho'\frac{2\pi}{\Lambda} . \sin 2\pi\frac{x}{\Lambda} + \rho'\rho_1\frac{2\pi}{\Lambda} . \cos 2\pi\frac{x}{L} . \sin 2\pi\frac{x}{\Lambda}.$$

For a layer of thickness dx at a depth x the amplitude of scattering is proportional to $\rho_x\,dx$ and the phase angle is $2\pi.2x\sin\theta/\lambda$. Thus the amplitude of scattering from the thin layer at level x is proportional to

$$\rho_x\,dx.\cos 2\pi 2x.\sin\theta/\lambda.$$

On inserting the above value of ρ_x we obtain terms involving the products:

$$(a)\ \cos 2\pi\frac{x}{L}.\cos 2\pi.\frac{2x\sin\theta}{\lambda},$$

$$(b)\ \sin 2\pi\frac{x}{\Lambda}.\cos 2\pi.\frac{2x\sin\theta}{\lambda},$$

$$(c)\ \cos 2\pi\frac{x}{\Lambda}.\sin 2\pi\frac{x}{L}.\cos 2\pi.\frac{2x\sin\theta}{\lambda},$$

and

$$(d)\ \cos 2\pi\frac{x}{L}.\sin 2\pi\frac{x}{\Lambda}.\cos 2\pi.\frac{2x\sin\theta}{\lambda}.$$

On integration from $x = 0$ to $x = \Lambda$ a non-zero result will only be obtained when $2\sin\theta/\lambda$ is equal to one of the following values:

$$(a)\ 1/L,\quad (b)\ 1/\Lambda,\quad (c)\ (1/L+1/\Lambda),\quad (d)\ (1/L-1/\Lambda),$$

and Λ is an integral multiple of L.

It is assumed that Λ is large compared with L so that the condition (*b*) only applies to X-rays suffering a very small deviation.

The θ-value corresponding to $\sin^{-1}\lambda/2L$ is the value discussed in section 1:2.2.1 and corresponds to the main X-ray reflection. The θ-values corresponding to conditions (*c*) and (*d*) give the satellites on either side of the main reflection. The amplitude of the waves corresponding to these satellites is given by the product $\rho\rho'$, i.e. the product of the amplitude of two sinusoidal density distributions.

These results can be conveniently expressed in terms of distances in reciprocal space. If we denote all reciprocal distances with an asterisk and choose our constant of reciprocity† as λ, then

$$\lambda/L = 2\sin\theta = L^*, \tag{1.2}$$

$$\lambda/L \pm \lambda/\Lambda = 2\sin\theta' = L^* \pm \Lambda^*.$$

The directions of L^* and Λ^* are perpendicular to the reflecting surface.

† Later in this book it is more convenient in theoretical discussion to take this constant as unity.

Thus, when the direction of the incident X-rays corresponds to a value θ' given by equation (1.2), a reflection due to the elastic waves occurs. The corresponding reciprocal points occur at distances $\pm\Lambda^*$ on either side of the usual reciprocal lattice point.

1:2.2.3 *Diffraction from a lattice perturbed by long waves*

The results of section 1:2.2.2 can be generalized and applied to a three-dimensional lattice. The distribution of electrons over the unit cell can be resolved, according to the usual Fourier analysis, into a series of sinusoidal distributions. We may write

$$\rho_0(\mathbf{r}) = \sum_m A(m)\cos 2\pi\{\mathbf{R}_m^*.\mathbf{r}+\alpha_m\}, \tag{1.3}$$

where $\rho_0(\mathbf{r})$ is the electron density at a point in the unit cell at the end of the vector $\mathbf{r}$ drawn from the origin;

$A(m)$ is the amplitude of the electron-density wave associated with a reciprocal point m;

$\mathbf{R}_m^*$ is the vector in reciprocal space (rel-vector) joining the origin and the reciprocal lattice point (relp) m; and

α_m is the phase angle of the electron-density wave, which in the present work will be put equal to zero, since only centro-symmetrical crystals will be under consideration.

A single wave is supposed to perturb the electron distribution in the crystal by a longitudinal displacement in the direction of the wave normal. We may write for the displacement $\mathbf{u}$ due to this wave,

$$\mathbf{u} = \boldsymbol{\xi} \cos 2\pi(\mathbf{K}^*.\mathbf{r}-\nu t), \tag{1.4}$$

where $\boldsymbol{\xi}$ is the amplitude vector and $\mathbf{K}^*$ is a vector in reciprocal space directed along the normal to the wave and of length $1/\Lambda$. In section 1:2.2.2 we saw that there are two effects due to the long wave—one due to the general displacement of matter which it causes and the other due to the change of density caused by the compression or rarefaction in the wave. We may generalize our previous result by expressing the perturbed density, $\rho_0(\mathbf{r})'$, as the product of two terms corresponding respectively to the general displacement and the compression, namely,

$$\rho_0(\mathbf{r})' = \left(\rho_0(\mathbf{r})-u_1\frac{\partial\rho_0}{\partial x_1}-u_2\frac{\partial\rho_0}{\partial x_2}-u_3\frac{\partial\rho_0}{\partial x_3}\right).\left(1-\frac{\partial u_1}{\partial x_1}-\frac{\partial u_2}{\partial x_2}-\frac{\partial u_3}{\partial x_3}\right),$$

where u_1, u_2, u_3 are the components of $\mathbf{u}$.
Neglecting products of small quantities, we obtain

$$\rho_0(\mathbf{r})' = \rho_0(\mathbf{r})-\frac{\partial}{\partial x_1}(\rho_0 u_1)-\frac{\partial}{\partial x_2}(\rho_0 u_2)-\frac{\partial}{\partial x_3}(\rho_0 u_3). \tag{1.5}$$

From equations (1.3) and (1.4) we obtain

$$\rho_0 u_1 = \sum_m A(m)\cos 2\pi(\mathbf{R}_m^*.\mathbf{r})\xi_1 \cos 2\pi(\mathbf{K}^*.\mathbf{r}-\nu t)$$

$$= \tfrac{1}{2}\sum_m A(m)\xi_1[\cos 2\pi(\mathbf{R}_m^*+\mathbf{K}^*.\mathbf{r}-\nu t)+\cos 2\pi(\mathbf{R}_m^*-\mathbf{K}^*.\mathbf{r}+\nu t)],$$

where ξ_1 is the component of ξ parallel to axis X_1. Differentiating this expression we have

$$\frac{\partial}{\partial x_1}(\rho_0 u_1) = \tfrac{1}{2}\sum_m A(m)\xi_1\Big[-2\pi\frac{\partial}{\partial x_1}(\mathbf{R}_m^*+\mathbf{K}^*.\mathbf{r})\sin 2\pi(\mathbf{R}_m^*+\mathbf{K}^*.\mathbf{r}-\nu t)-$$

$$-2\pi\frac{\partial}{\partial x_1}(\mathbf{R}_m^*-\mathbf{K}^*.\mathbf{r})\sin 2\pi(\mathbf{R}_m^*-\mathbf{K}^*.\mathbf{r}+\nu t)\Big]. \quad (1.6)$$

No matter what the system of symmetry to which the crystal belongs, we may write

$$\mathbf{R}_m^*+\mathbf{K}^* = \mathbf{H}^* = h_1\mathbf{n}^*+h_2\mathbf{o}^*+h_3\mathbf{p}^*,$$

where $\mathbf{n}^*$, $\mathbf{o}^*$, $\mathbf{p}^*$ are the sides of the 'reference' cubic unit cell in reciprocal space corresponding to the cubic unit cell in Bravais space having sides **n**, **o**, **p**, and h_1, h_2, h_3 are any integral or non-integral coefficients (Fig. 1.10).

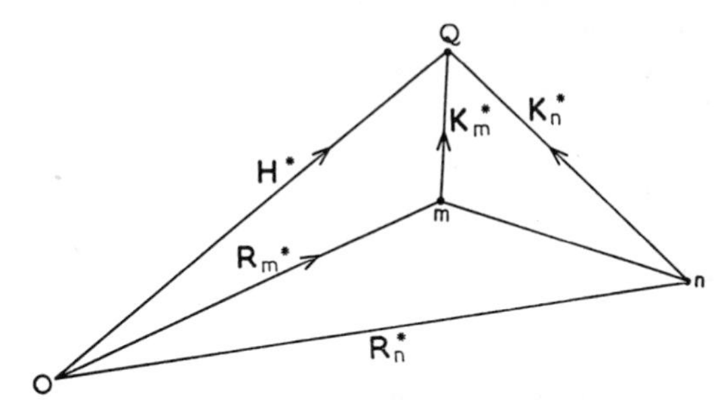

FIG. 1.10 Diagram illustrating the connection between various vectors in reciprocal space.

We may also write

$$\mathbf{r} = x_1\mathbf{n}+x_2\mathbf{o}+x_3\mathbf{p},$$

where x_1, x_2, x_3 are any integral or non-integral coefficients. Then,

$$\mathbf{R}_m^*+\mathbf{K}^*.\mathbf{r} = h_1x_1+h_2x_2+h_3x_3.$$

Now

$$\frac{\partial}{\partial x_1}(\mathbf{R}_m^*+\mathbf{K}^*.\mathbf{r}) = h_1, \qquad \frac{\partial}{\partial x_2}(\mathbf{R}_m^*+\mathbf{K}^*.\mathbf{r}) = h_2, \qquad \text{etc.}$$

Thus the sum of the three partial differential terms in equation (1.5) has a factor $\xi_1h_1+\xi_2h_2+\xi_3h_3$ which is equal to $\xi.\mathbf{R}_m^*+\mathbf{K}^*$. Similarly, the second half of expression (1.6) leads to a term $\xi.\mathbf{R}_m^*-\mathbf{K}^*$. Thus the full evaluation of equation (1.5) gives

$$\rho_0(\mathbf{r})' = \rho_0(\mathbf{r})+\pi \sum_m A(m)[(\xi.\mathbf{R}_m^*+\mathbf{K}^*).\sin 2\pi(\mathbf{R}_m^*+\mathbf{K}^*.\mathbf{r}-\nu t)+ \\ +(\xi.\mathbf{R}_m^*-\mathbf{K}^*).\sin 2\pi(\mathbf{R}_m^*-\mathbf{K}^*.\mathbf{r}+\nu t)]. \quad (1.7)$$

The term $\rho_0(\mathbf{r})$ corresponds to the density distribution which is responsible for the Bragg reflection. The second term gives the diffuse scattering due to the elastic waves. This distribution is sinusoidal and $\mathbf{R}^*+\mathbf{K}^*$ or $\mathbf{R}_m^*-\mathbf{K}^*$ have replaced $\mathbf{R}_m$ in the expression (1.3). Thus we see that equation (1.7) shows that there is an additional electron density associated with the disturbed lattice which corresponds to reciprocal points displaced by the vector $\mathbf{K}^*$ on either side of any reciprocal lattice point (relp) defined by $\mathbf{R}^*$. This gives rise to the amplitude $\pi A(m)\xi.\mathbf{H}^*$ for a given value of m, i.e. near to a particular relp defined by m. The intensity of this scattering is thus proportional to $\pi^2 A(m)^2(\xi.\mathbf{H}^*)^2$. Now there are always waves of the same wavelength and velocity travelling in opposite directions so that the contributions from these two trains of waves to the points $(\mathbf{R}^*+\mathbf{K}^*)$ and $(\mathbf{R}^*-\mathbf{K}^*)$ just double the intensity of scattering, at either point. If we express the scattering from one unit cell of the crystal as a ratio to that given by one classical electron scattering under the same conditions, we may replace the amplitude of the electron-density wave $A(m)$ by the modulus of the structure amplitude $|F_{hkl}|$, where hkl are the indices of the point m. The appropriate value of $|F_{hkl}|$ is that for the temperature of observation, after applying the Hönl correction (see James, 1948, p. 180) and will be denoted $|F_T|$. Thus we may write the intensity $I(\mathbf{H}^*)$ of the first order thermal diffuse scattering due to any one wave of amplitude vector ξ according to the equation,

$$I(\mathbf{H}^*) = 2\pi^2|F_T|^2(\xi.\mathbf{H}^*)^2 \quad (1.8)$$

Associated with any wave vector $\mathbf{K}^*$ there are three amplitude vectors, denoted $\xi_{(1)}$, $\xi_{(2)}$, and $\xi_{(3)}$ respectively. $\xi_{(1)}$ corresponds to the almost longitudinal wave and $\xi_{(2)}$ and $\xi_{(3)}$ to the almost transverse waves associated with waves having a given wave normal. Each of these three waves contributes independently to the diffuse scattering and we must therefore add their intensities. The final expression is given by

$$I(\mathbf{H}^*) = 2\pi^2|F_T|^2 \sum_{i=1}^{3} (\xi_{(i)}.\mathbf{H}^*)^2. \quad (1.9)$$

$I(\mathbf{H}^*)$ is known as the first order Diffuse Scattering Power and is denoted D_1. The units in which D_1 is expressed are the same as those of $|F_{hkl}|^2$.

When the volume element of reciprocal space giving rise to the diffuse scattering is not close to a particular relp, as is the case for studies of order-disorder and molecular shape, it is important to take account of the fact that a large number of elastic waves may be concerned in the scattering. In Fig. 1.10 any two relps, denoted m and n respectively, are joined to the point Q, from which the diffuse scattering is arising, by vectors $\mathbf{K}_m^*$ and $\mathbf{K}_n^*$. It can be seen from the diagram that

$$\mathbf{H}^* = \mathbf{R}_m^* + \mathbf{K}_m^* = \mathbf{R}_n^* + \mathbf{K}_n^* = \text{etc.}$$

All the waves corresponding to $\mathbf{K}_m^*$, $\mathbf{K}_n^*$, etc., contribute to the scattering.

An important effect associated with the fact that the elastic waves are travelling through the crystal with a finite velocity is that there is a Doppler change of frequency of the X-rays on reflection. This change of frequency is so small that it cannot be detected by a normal X-ray spectrometer but it is great enough to make the diffusely scattered radiation incoherent. In the calculation of intensity of the diffusely scattered radiation we may always sum the squares of the amplitudes of the separately scattered waves. This is implicit in equation (1.9) and applies equally to second and higher order diffuse scattering.

1:2.3 *The reciprocal lattice corresponding to waves travelling in a crystal*

Just as standing waves on a rope which is fixed at both ends must have wavelengths which bear a simple relation to the length of the rope, so the elastic waves in a crystal are limited by the boundary conditions. In all the work under discussion the wavelengths of the elastic waves lie in the range 40–200 Å, and, therefore, the dimensions of a crystal which can be used in the experiments described here are practically infinite in comparison with the wavelength of the elastic waves. Standing waves in an infinite lattice cannot produce any net expansion which is not somewhere else compensated by an equal contraction. In other words, the distance between the opposite faces of a crystalline block must be a whole multiple of the wavelength of any standing elastic wave (Born and Karman, 1912, 1913). This sets a limit to the longest wavelengths which can occur but this limit is of theoretical rather than practical importance since the energy associated in thermal motion with the wavelengths comparable with the size of the crystal is small in comparison with the energy of the waves of shorter wavelength.

Another limit to the range of wavelengths found among the standing waves is due to the fact that the crystal is built on a lattice. A consequence of this is that the minimum possible wavelength is twice the length of the side of the unit cell. A one-dimensional illustration will serve to show how the lower limit to the possible elastic wavelengths arises. In Fig. 1.11 a one-dimensional crystal having twelve unit cells is represented by the horizontal lines. The lattice points are shown displaced by waves which, in the drawing, are shown as transverse but in the one-dimensional crystal must be regarded as longitudinal. Within the length of this crystal are shown, in descending order, the displacements corresponding to one up to eight complete waves within the

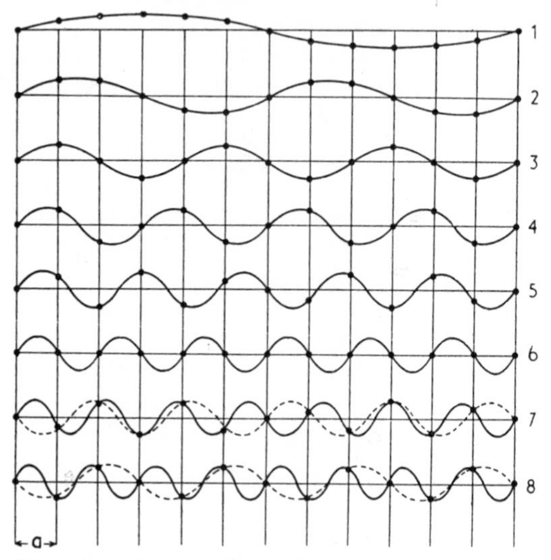

FIG. 1.11 Diagram illustrating the atomic displacements due to harmonics of a given fundamental wave motion.

twelve unit cells. The limiting case is that in which there are six waves of length twice as great as that of the cell side. If the wavelength is made smaller than this, as is shown in the diagrams with seven and eight waves respectively, the displacements also correspond to wavelengths which are greater than $2a$. The waves for line 7 in Fig. 1.11 are shown full for the seven waves and dotted for the equivalent five waves. The full line in the last line corresponds to eight waves in the length of the crystal and the dotted line to the equivalent four waves. The equivalence may be shown generally as follows.

The displacement, δ, of any point in terms of the distance, x, from the left-hand edge is given by

$$\delta = A \sin 2\pi \frac{x}{a} \cdot \frac{n}{12},$$

where n is a whole number, corresponding to the number of complete waves within the length $12a$.

If we put

$$m = 12 - n$$

we may write

$$\delta = A \sin 2\pi \frac{x}{a}\left(1 - \frac{m}{12}\right)$$

$$= -A \sin 2\pi \frac{x}{a} \cdot \frac{m}{12},$$

since x is always a multiple of a at any lattice point. Thus when n exceeds 6, i.e. when the elastic wavelength is shorter than $2a$, the displacements of the lattice points are always equivalent to those of a value of m equal to $12-n$.

The above discussion relates to a lattice consisting of identical point-like atoms. If at each lattice point there is a molecule consisting of a number of atoms, the equivalence between waves of wavelength less than and greater than $2a$ disappears. This may be seen by reference to Fig. 1.11 where the dotted and full lines in the last two horizontal lines do not overlap. The displacements of the atoms in molecules whose centres are situated at the lattice points will therefore be different for the waves corresponding to the full and dotted curves respectively.

We can now summarize all the information concerning the connection between the elastic waves and the reciprocal lattice. Since the boundary conditions require that a whole number of wavelengths shall lie between the bounding faces of the crystal, the elastic wave fronts must all be parallel to lattice planes. In other words, the wave vectors $\mathbf{K}^*$ must all be in directions parallel to the vectors joining reciprocal lattice points to the origin (rel-vectors), i.e. to $\mathbf{R}_m^*$. Moreover, the lengths $\mathbf{K}^*$ must all be sub-multiples of the corresponding $\mathbf{R}_m^*$. It is, therefore, possible to construct a so-called *elastic reciprocal lattice* in which each point represents one wave, its distance from the origin being K^*. The linear reciprocal lattice shown in Fig. 1.12 corresponds to the waves in the linear lattice of Fig. 1.11. The lower row of numbers

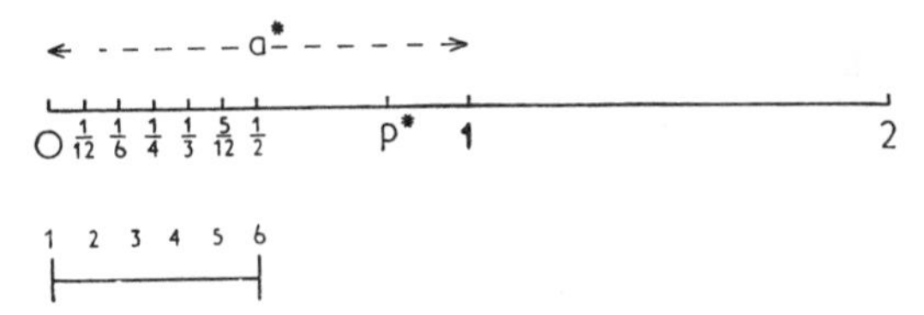

FIG. 1.12 Diagram illustrating the reciprocal equivalent of the wave motions figures in Fig. 1.11.

in Fig. 1.12 corresponds to the numbers on the right-hand side of Fig. 1.11. It will be seen that the reciprocal points occur at distances from the origin equal to a^*/K^*, i.e. 1/12, 1/6, 1/4, 1/3, 5/12, 1/2 of the unit reciprocal distance a^*. These reciprocal points lie in a lattice, and, although not all points of this lattice are occupied, the reciprocal elastic lattice can be constructed with a repeat distance of $a^*/12$. So long as only single atoms occupy the lattice points the elastic reciprocal points do not extend further from 0 than $a^*/2$. This may also be expressed as follows. If any wave motion is represented by a point in reciprocal space further from 0 than $a^*/2$, say at p^*, it can equally well be represented by a point distant a^*-p^* from 0, i.e. between 0 and $a^*/2$. This limitation requires qualification when there are molecules at the lattice points.

These results can be generalized for a three-dimensional lattice as follows. Let us assume that the crystal is in the form of a parallelepiped of sides of length N_1a_1, N_2a_2, and N_3a_3 respectively. (It can be shown that the actual shape of the crystal is irrelevant when it is as large as those which are used in practice.) Let us consider an elastic wave parallel to a lattice plane (hkl). In order to satisfy the boundary conditions that within the crystal there shall be a whole number of wavelengths we can state the following. The number of lattice planes of indices (hkl) which cut the sides of the parallelepiped are N_1h, N_2k, N_3l respectively. If the wavelength of the elastic wave is n times that of the spacing of the lattice planes, then the number of elastic wave fronts cutting the three edges of the crystal block is N_1h/n, N_2k/n, N_3l/n. The reciprocal vector $\mathbf{R}^*$ is given by

$$\mathbf{R}^* = h\mathbf{a}_1^* + k\mathbf{a}_2^* + l\mathbf{a}_3^*,$$

and the reciprocal vector of the elastic waves, K^*, is given by

$$\mathbf{K}^* = \frac{1}{n}\mathbf{R}^* = \frac{h}{n}\mathbf{a}_1^* + \frac{k}{n}\mathbf{a}_2^* + \frac{l}{n}\mathbf{a}_3^*.$$

Thus the elastic waves may be represented by a reciprocal lattice of a scale $1/n$ of that of the normal reciprocal lattice. The upper limiting value of n is $N_1/2$, $N_2/2$, $N_3/2$ along the three sides of the parallelepiped. The lower limiting value of n for a lattice having one atom per cell is 2. Thus an elastic reciprocal lattice appropriate for a lattice of one atom per cell can be constructed consisting of unit cells of sides $2\mathbf{a}_1^*/N_1$, $2\mathbf{a}_2^*/N_2$, $2\mathbf{a}_3^*/N_3$, and every possible wave is represented by one or other of these reciprocal points. The total number of these elastic reciprocal points is thus $N_1N_2N_3$, i.e. the same as the total number of atoms in the crystal block, and each point represents one longitudinal and two transverse waves.

All the points in an elastic lattice relating to a unit cell having only one atom are contained within what is known as the *First Brillouin Zone*. From a geometrical point of view this is defined as follows. The rel-vectors are drawn from the origin to the nearest relps. Planes are now drawn normal to each of these rel-vectors and passing through their mid-points. The space enclosed by these intersecting planes is known as the *First Brillouin Zone*. The second such zone may be similarly constructed using the reciprocal points next nearest to the origin. An exactly similar zone can be constructed round each relp and we shall make much use of such zones in evaluating the intensities of diffuse reflections.

A wave represented by a point P (Fig. 1.13) outside the first Brillouin Zone can equally well be represented by a point, P', within that zone. Suppose $\mathbf{K}^*$ is the wave vector of point P and A is the nearest relp. Then, if

$$\mathbf{AP} = \mathbf{K}^*,$$

$$\mathbf{H}^* = \mathbf{R}^* + \mathbf{K}^*.$$

Any point on the Bravais lattice is joined to the origin by a vector **r**. The displacement at time zero, $\mathbf{u}_r$, of this point by the wave is given by

$$\mathbf{u}_r = \boldsymbol{\xi} \cos 2\pi \mathbf{K}^* . \mathbf{r},$$

where $\boldsymbol{\xi}$ is the amplitude vector of the wave.

Now, $$\cos 2\pi \mathbf{K}^* . \mathbf{r} = \cos 2\pi (\mathbf{R}^* + \mathbf{K}^*) . \mathbf{r}$$

since the product $\mathbf{R}^* . \mathbf{r}$ is a whole number.

Thus the wave having **OP** as its wave vector produces everywhere the same displacement as the wave having **OP**′ as its wave vector. The point P' lies within the first Brillouin Zone, and will generally be used to represent the wave motion having a wave normal corresponding to **K***.

1:2.4 *Second and higher order diffuse scattering*

In the theoretical analysis given up to this point we have evaluated the diffuse scattering due to the wave vector **K*** which connects any arbitrarily chosen region of reciprocal space P (Fig. 1.13) with its nearest relp, A. The

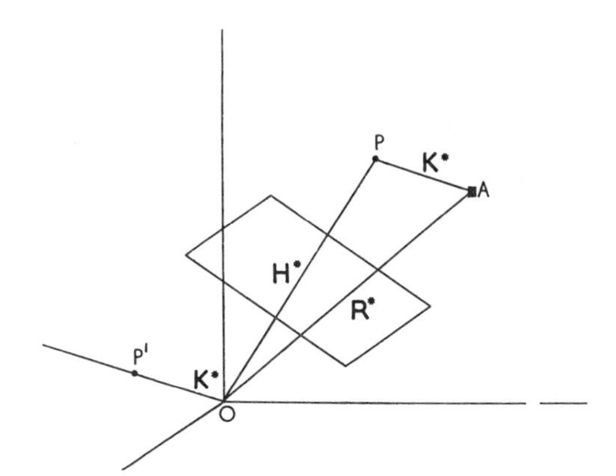

FIG. 1.13 Diagram illustrating the equivalence of reciprocal vectors in different Brillouin Zones.

diffuse scattering associated with the wave normal **K*** is called first order because only one wave normal is involved. A small contribution to the diffuse scattering associated with the point P arises from doubly and trebly scattered waves. For example, the first order diffuse scattering associated with the point Q (Fig. 1.14) arises by the superposition of the long wavelength waves corresponding to the wave normal AQ on the structural periodicity corresponding to OA. Starting with the periodicity corresponding to OQ and

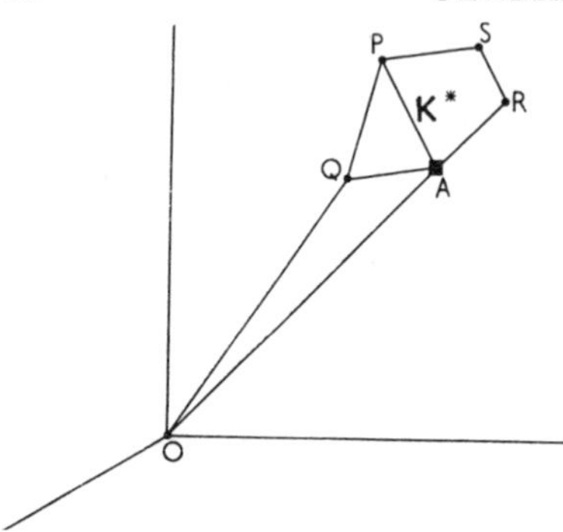

FIG. 1.14 Diagram showing the vectors involved in first, second, and third order diffuse scattering.

combining it with the wave displacements corresponding to the wave normal *QP*, we obtain the resultant periodicity for the doubly scattered waves corresponding to *OP*. There are, of course, a very large number of points such as *Q* on the elastic reciprocal lattice and an integration over all of them must be made in evaluating the total contribution due to the second order diffuse scattering. Similarly, a third order diffuse scattering can arise from the combination of the effects of reciprocal points such as *R* and *S* (Fig. 1.14). Since the vector relation

$$\mathbf{AR}+\mathbf{RS}+\mathbf{SP} = \mathbf{AP}$$

is obeyed, the X-rays successively scattered by waves having wave vectors **AR, RS,** and **SP** must travel in the same direction as those scattered only once by the wave having **AP** as wave vector.

Even the second order diffuse scattering can often be neglected in comparison with the first order scattering. The third order scattering hardly ever has sufficient intensity to make an appreciable contribution to the observed effect. The evaluation of the intensity of the second order diffuse scattering will be given in later sections. The second order Diffuse Scattering Power is denoted D_2 and this symbol is exactly analogous to D_1 in expressions for the scattered intensity.

1:3 Experimental methods

As in so many branches of crystal physics involving the use of X-rays, there are two methods by which the experimental data may be obtained. One method uses photographic plates and films, the other uses Geiger counters or other ionization or scintillation detectors. It is not possible to say that the photographic method is necessarily better than the ionization method, nor is the converse statement possible. Under different circumstances and for different purposes each of these methods has its own proper sphere of application. The possibilities of the laboratory and the kind of other apparatus available also affect the method to be employed. We may begin by giving a rough outline of the circumstances under which the photographic method can be employed.

In the first place the apparatus required for the photographic method is, generally speaking, simpler than that required for the ionization method. The photographs are usually taken with filtered radiation, though, where the intensity of the incident X-ray beam is great enough, there are many advantages in using crystal-reflected radiation. However, much can be done using only filtered radiation. Secondly, the cameras employed need only be of a simple kind, though again, if more complicated cameras are available, they can be used with advantage. Then the study of the photographic films involves some form of microdensitometer, which measures, not the transparency of the film but the density, to which the incident X-ray intensity is proportional. Altogether, the equipment involved for making photographic measurements is no greater than that normally used in connection with the usual structural investigations employing a photographic method.

The one great advantage which the photographic method has over the ionization method is that the length of time during which observations may be made may extend up to a hundred hours, whereas with the ionization method any single observation must be completed in a few minutes, perhaps not more than fifteen, or it cannot be made at all. That means, in fact, that the photographic method can study far weaker reflections, and far weaker parts of the pattern, than could be studied, even under the best conditions, with the ionization method.

Another great advantage of the photographic method is that, generally speaking, the steadiness of the output of the X-rays is of no great importance. The photograph extends over a long time and the output of the X-rays naturally fluctuates during this time for reasons which vary from place to place. Sometimes the voltage of the mains is not constant, sometimes the temperature of the room may fluctuate considerably and this may give rise to relative changes in the position of the camera and the X-ray tube; but even with all these possible variations the photographic method gives a result which is quite satisfactory, because all parts of the pattern are being produced all the time. The only exception to this statement is when the double crystal method is used for the purpose of calibrating the intensity of the diffuse reflections. The standard scatterer, whatever it may be, is not in the beam when the crystal under investigation is in the beam, and a different sample of the X-ray beam is used, so that the results may depend upon the fluctuations in the output of the X-ray beam. However, for many purposes this is an unnecessary refinement and it is true that the observations are not seriously dependent on the steadiness of the output of X-rays or the sensitivity of the recording film. The results are dependent, of course, on the uniformity of the processing of the film. If the film is not uniformly developed, or if for reasons connected with the thickness of the emulsion or the distribution of the grains within the emulsion, the sensitivity of one part differs from that of another, then errors occur in the results.

The great merit of the photographic method is that a considerable part of the total pattern of diffuse reflection is recorded at the same time and generally speaking it is advantageous to begin an investigation by using photographic methods and finish it by using ionization methods. The photographic method is valuable for indicating what features of the pattern must be carefully studied, and the ionization method is more useful or perhaps more accurate in obtaining the quantitative data on the features of the pattern which are known to be interesting from the photographic study.

The ionization method gives essentially a reading of the intensity of the diffuse reflections in a particular direction for a given setting of the crystal. It is possible to decide beforehand that a certain limited number of directions of observation should be employed, to go through these in order, and to register the intensity of the diffuse reflection travelling from the crystal in those directions. This is a convenient method when only a limited number of directions need to be considered, as for instance in the study of the elastic properties of crystals. But it is a tedious and extremely laborious method when the number of directions which has to be considered is large, as, for instance, whenever a three-dimensional survey of the diffuse scattering in the whole of a volume of reciprocal space must be determined.

The question of accuracy of observation is also involved in the choice of method. It is generally agreed that by photographic means the intensity of the X-ray beam incident on any given area of the film can be determined with an accuracy of a few per cent. The precise value depends upon the apparatus employed and the care taken to standardize the conditions of the experiment, but unless special precautions are taken it is difficult to attain an accuracy greater than 5 per cent.

The ionization method depends for its accuracy primarily upon the number of counts which can be made in any given direction. Owing to random fluctuations the number of counts which must be made in order to attain the accuracy of 3 per cent. is 1,000, and for an accuracy of 1 per cent. 10,000. Provided the intensity of the X-ray beams is great enough, so that 10,000 counts can be registered in a reasonable time, then there is no difficulty in attaining an accuracy of 1 per cent., and it is in this way that a greater accuracy can be obtained by the ionization method than by the photographic method. Against this must be set a geometrical difference. The photographic film is capable of registering the intensity of the diffusely reflected X-rays over a small area of the order of $0{\cdot}1 \times 0{\cdot}1$ mm. Under practical conditions this subtends a smaller angle at any point on the surface of the crystal than the corresponding windows of a Geiger counter or other detector which may be used in the ionization method. The angular resolution which can be attained by the photographic method is considerably higher than that which can be attained by the ionization method.

1:3.1 *Photographic method*

1:3.1.1 *Goniometers*

For many purposes involving single crystals a standard oscillation goniometer provided with a cylindrical or hemi-cylindrical camera of radius 5·73 cm is suitable. A radius less than this is too small to give adequate angular resolution, a radius larger than this increases the length of exposure without necessarily giving much in the way of increased resolution, though for certain purposes it may be worth while to employ a camera of a larger radius. For powder investigations a focusing camera of the Guinier type is, generally speaking, very valuable, particularly when the radiation can be crystal-reflected so that the information is not distorted by the presence of radiation other than the characteristic.

It may sometimes be desirable to impress upon the photograph the reflections from some other crystal so that a standard of intensity may be placed on the film at the same time as the diffuse reflection is being recorded. A modified Weissenberg two-crystal instrument was used for this purpose by Hoerni and Wooster (1952*b*). The goniometer was arranged so that after fifteen minutes the crystal under test was moved out of the X-ray beam and a standard crystal was moved into the beam and oscillated once through the Bragg reflecting position. The film was similarly moved so that the two reflections did not overlap and a screen was, of course, employed to ensure that the film was protected at the appropriate place while the exposure was being made elsewhere. In this way it was possible to register standard reflections and the reflections under investigation on the same piece of film, not quite at the same time, but at points in time, which gave a representative sample of the intensity of the incident X-rays. A full description of many types of goniometers and diffractometers has been given by Umanskii (1960).

1:3.1.2 *Measurements on the photographs*

The measurements on the photographs are of two kinds. In the first place measurements of position of the reflected X-rays are important in the interpretation, and secondly the intensity, especially the relative intensity, at different points over the region must be measured. The measurement of position is best achieved by impressing upon the photograph some standard reflection, which may often be a characteristic reflection from some plane of the crystal under investigation. Such a reflection may be impressed on the film itself if it does not interfere with the pattern of the diffuse reflection or it may be impressed at a different part of the film if it does so interfere, provided that the film can be moved along between the recording of the diffuse reflection and the recording of the characteristic reflection.

In order to obtain sufficient precision in the determination of the position of points on the film it is necessary to use collimators which are as small as are consistent with obtaining exposures in a reasonable time. Usually a collimator with three apertures, $\frac{1}{2}$, $\frac{1}{4}$, $\frac{1}{2}$ mm respectively, is convenient though on some occasions it may be necessary to use smaller diameter apertures than these.

The measurement of the intensity of the diffuse pattern is almost impossible without automatic recording. When a hand-operated instrument is used the operator tends to select a limited number of points and the fluctuations in the intensity of the photographic record from point to point are usually so large that a representative determination of the intensity over the diffuse spot is not attained. On an automatic record the fluctuations over a considerable area can be seen and allowance can be made for them in drawing an average curve through the fluctuations of the record.

The following is a short description of the microdensitometer† which was

† Made by Crystal Structures Ltd., Cambridge.

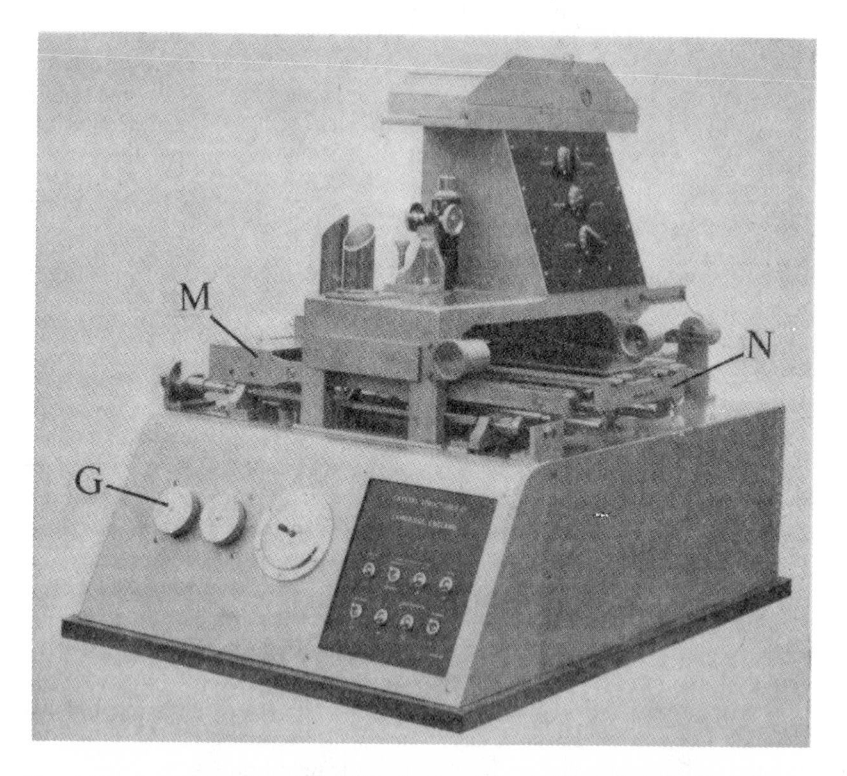

FIG. 1.15. A microdensitometer suitable for the study of diffuse photographic patterns. *M*, *N* are film carriages giving movements along two perpendicular directions. *G*, controls of 6-speed gear boxes, one of which drives *M* and the other *N*.

developed for the measurement of both the position of points on a diffuse record and also of the X-ray intensity corresponding to the reflections (Wooster, 1960*b*). A photograph of the instrument is given in Fig. 1.15 and a schematic diagram of the component parts in Fig. 1.16.

Light beams from the single lamp *L*, Fig. 1.16, travel by two routes to a single photocell *P*. The beams are interrupted by a rotating shutter which simultaneously cuts off the light along one path as it permits it to pass along the other. The condensers C_1, C_2 focus the straight filament of the lamp on to the absorbing layer —the optical wedge on the left, the film under test on the right. The objectives O_1, O_2, supplemented when necessary by the projecting eyepieces E_1, E_2, produce an enlarged image of the lamp filament in the plane of the apertures A_1, A_2, and the lenses and prisms give overlapping images on the sensitive surface of the photocell *P*. The motor driving the shutter also drives an a.c. generator, the output of which is necessarily in step with that of the photocell.

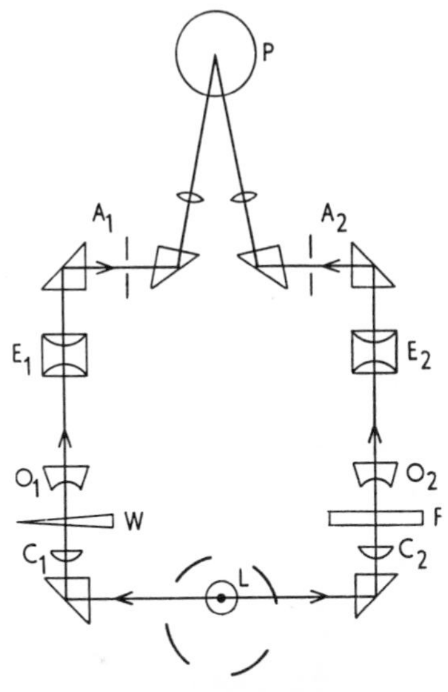

FIG. 1.16 Diagrammatic representation of the components of the optical system of the microdensitometer. Illustrated in Fig. 1.15.

The microdensitometer is used throughout as a nul instrument. The photographic wedge *W* is moved by a servo-motor along its length until the intensity of light passing through it balances that passing through *F*, the film under test. When the intensities of the two light beams are equal there is no alternating output from the photocell. When the left-hand beam is stronger than the right, there is an alternating output having a phase which is 90° in advance (or retard) with respect to that from the a.c. generator. When the right-hand beam is the stronger the phase of the photocell output is 90° in retard (or advance). The direction of rotation of the servo-motor driving the wedge makes use of this phase change. The frequency of this a.c. is 190 c/s in order to avoid spurious effects due to the mains frequency of 50 c/s which heats the lamp filament.

The wedge movement is coupled to a potentiometer and the output from this is applied to the pen-recorder. Synchronous motors are used both on the drive for the film under test and in the chart motor, so that the necessary relation between the two movements is maintained. The speed of translation of the film under test can be changed by 32:1 in a series of six 2:1 steps simply by rotating a handle G (Fig. 1.15). As the pen-recorder has a further range of speeds of 4:1 it is possible to make a very rapid, approximate survey of a film, or a slow, accurate investigation. Diffuse reflection photographs are usually scanned at a speed of 0·5 mm/min. The film to be surveyed is supported on the carriage, M, Fig. 1.15, by means of which it can be traversed at a uniform speed under the microscope O_2E_2. This carriage is provided with a millimetre scale by means of which the relative positions of points along the film in the direction of movement can be read off. A second carriage, N, is mounted on top of the first and this enables the film to be moved to any desired position in a direction perpendicular to the movement of the carriage M. The position of the second carriage can also be read off on a millimetre scale. With these two perpendicular movements of the film it is possible to survey a diffuse spot by making a succession of scans along parallel equidistant lines which are 0·5 mm or 1·0 mm apart.

The diffuse spot usually covers an area of a few millimetres by a few millimetres though in some of the records involved in the determination of molecular shape the area to be studied is much larger and in fact covers practically the whole film, which may be 10 cm × 10 cm. But when a small area is to be investigated it is useful to inscribe on the film four lines which form a square round the area. The film is mounted on the stage of the microdensitometer so that it can be moved in two directions at right angles, which we denote x, y, respectively, and it is placed so that the top left-hand corner is on the axis of the scanning microscope. The film is now made to move at uniform speed parallel to the x-direction until the other side of the diffuse spot is reached. The film is then moved say half a millimetre along the y-direction and, starting again from the scribed line on the left-hand side of the spot, a second traverse is made in the x-direction. This process is repeated until the whole area of the diffuse spot is covered. On each record there are two marks indicating the positions of the scribed lines and in this way it is possible to combine all the information contained on the records. This is conveniently done by constructing a contour map in which the contours correspond to given levels of density.

The two ends of any one record must always occur sufficiently far from the diffuse spot in order to correspond to the general background. These ends of the record are joined by a straight line, AB, as shown in Fig. 1.17a. Lines are now drawn across the record parallel to AB and with a vertical separation of, say, 2 or 5 scale divisions of the record paper. The intersections of these lines with the record ACB are transferred to the base line. A contoured density map can be constructed using the same scale of magnification as is used on the chart. Thus in Fig. 1.17a the line AB corresponds to the line AB in Fig. 1.17b and it will be seen how the positions of the intersections of the contours with the horizontal lines in Fig. 1.17b enables the whole contoured map to be constructed.

When it is necessary to measure the distribution of diffuse intensity over the whole of an X-ray photograph, exactly the same method may be employed but, of course, the traverses must be made faster and the distance between parallel traverses must be made greater in order that the film may be covered in a

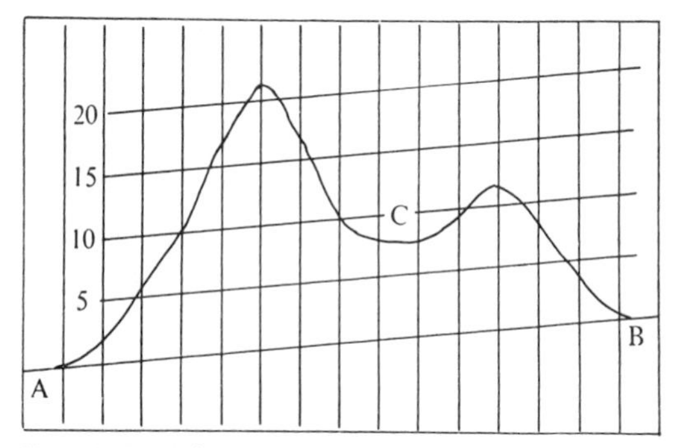

FIG. 1.17*a* Diagram showing how to find the positions along the microdensitometer trace, *ACB*, at which the density reaches given levels above the background density.

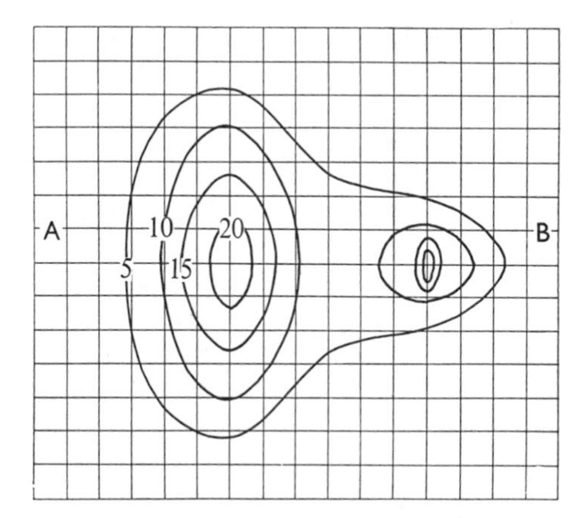

FIG. 1.17*b* Diagram illustrating the tracing out of a contour map of density by determining the positions at which given contours cross a line of scan such as *AB*.

reasonable time. The purpose of making such surveys over the whole film will usually be to plot out the transform of a molecule and this will, of course, be on a much larger scale than the diffuse spot which we have discussed above. It is, therefore, permissible to scan using larger distances between the parallel traces and also to scan at a greater rate. The method of constructing a contour map from the curves obtained on the microdensitometer is, in principle, exactly the same as that described above.

1:3.1.3 *Method of taking the photographs*

From an experimental point of view the simplest procedure in studying photographically the diffuse X-ray reflections from a single crystal is the following. The crystal is set up on a usual X-ray goniometer provided with a cylindrical or hemi-cylindrical camera of radius not less than 5·73 cm. The Bragg setting of the crystal defined by the (glancing) angle of incidence θ, which permits the characteristic X-rays to be reflected from a given relp, is found. This setting can be achieved in several ways but one simple method is as follows. The film is held in its camera by clips in such a way that the film can be easily moved by a centimetre in between each exposure. The crystal is set at angles which are believed to correspond to $\theta+1°$, $\theta+\frac{1}{2}°$, θ, $\theta-\frac{1}{2}°$, $\theta-1°$, and in each setting an exposure of a few seconds is given. After each exposure the film is moved so that the X-ray spots do not overlap and so that it is easy to relate the setting of the crystal to the corresponding spot on the film. From the developed photograph the setting that corresponds to θ can be seen. The accuracy of the setting can be improved by further tests of this kind with smaller angular intervals between successive settings. If a Geiger counter and auxiliary equipment is available this may also be used to find the crystal setting corresponding to the Bragg reflection. The crystal is then rotated from this θ-setting through an angle of between one and three degrees and the long exposure for the diffuse reflections is

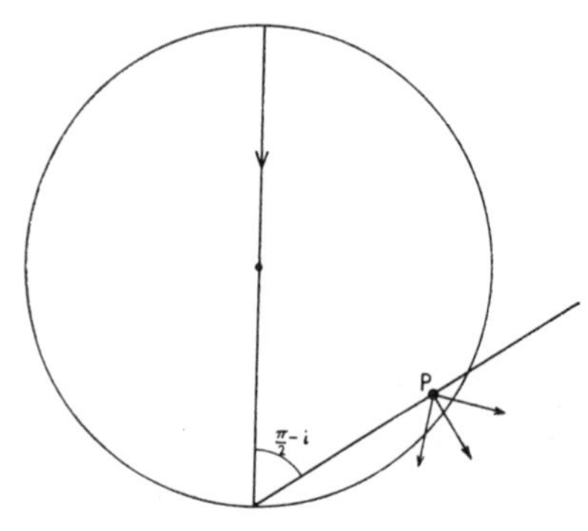

FIG. 1.18 Diagram showing the relp P within the reflecting sphere and the rekhas, represented by arrows, passing through it.

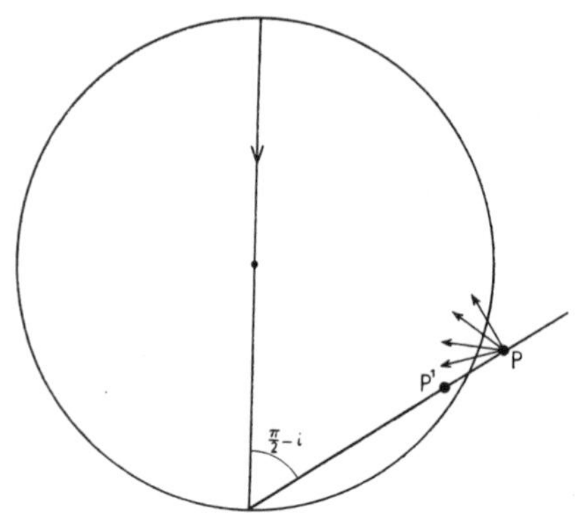

FIG. 1.19 Diagram showing the relp P outside the reflecting sphere and the corresponding rekhas giving diffuse scattering represented by arrows.

made. The length of exposure increases as the departure from the Bragg setting increases. If it requires a few seconds to produce an easily visible spot in the Bragg setting it will generally require some tens of hours to obtain a suitable density in the spot due to the thermal diffuse scattering, though other types of diffuse reflection often require only a few hours' exposure. If the angle of incidence of the X-rays in the setting for the diffuse reflection be denoted i, then i may be greater or less than θ by one to three degrees. If i is greater than θ, then the relp, P (Fig. 1.18), is inside the reflecting sphere. If, however, i is less than θ, then the relp, P, is outside the reflecting sphere (Fig. 1.19). In this setting the reflecting sphere passes between the relps P and P', corresponding respectively to the $K\alpha$- and $K\beta$-radiations. The diffuse reflection due to the $K\beta$-radiation can disturb the pattern due to the $K\alpha$-radiation unless a reflecting-crystal monochromator is used. If a β-filter is used it must be thick enough to reduce the intensity of the β-radiation to one per cent. or less of the α-radiation. The directions of the rekhas that can be examined are shown in Figs. 1.18 and 1.19 by arrows passing through the relp P. It will be seen that nearly the same angular range of rekhas can be studied in the two settings, since the thermal diffuse scattering is centrosymmetrical about P.

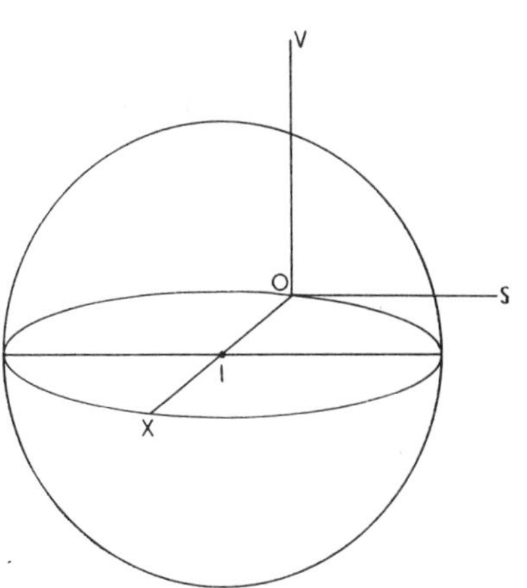

FIG. 1.20 Diagram showing the reference axes OX, OV, OS, which are commonly used in the interpretation of oscillation and other diffraction photographs.

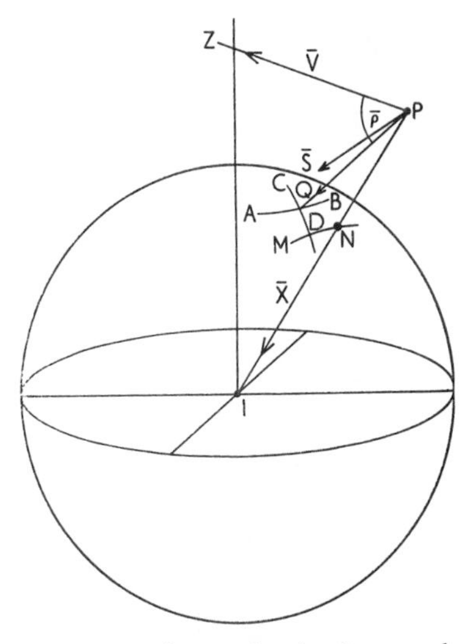

FIG. 1.21 Diagram showing the general reference axes $O\bar{X}$, $O\bar{V}$, $O\bar{S}$ in relation to the relp P and the reflecting sphere.

1:3.1.4 *The geometry of diffuse X-ray reflections*

In the interpretation of the normal X-ray diffraction photograph of a single crystal an axial system which is indicated in Fig. 1.20 is commonly used (Hoerni and Wooster, 1952*c*). This consists of a vertical axis, OV, a horizontal axis, OI, the direction IO being that of the incident X-rays, and a third axis, OS, at right angles to the other two axes. This axial system is convenient when the region of reciprocal space represented on any one photograph corresponds to a considerable area of the surface of the reflecting sphere shown in Fig. 1.20. When we are concerned only with regions of small volume round particular relps the axial system OV, OI, OS is not suitable and a more convenient axial system is shown in Fig. 1.21. The relp is denoted by the point P and the line PI represents the $\bar{X}$ axis. The vertical axis through I is denoted IZ. The axis $\bar{V}$ is normal to the axis $\bar{X}$ and lies in the vertical plane containing PI and IZ. At right-angles to $\bar{X}$ and $\bar{V}$ is a third axis denoted $\bar{S}$. The direction of a particular rekha is denoted by the line PQ, Q being the point on the surface of the reflecting sphere where the line PQ passes through it, and we employ two angles, $\bar{\rho}$, $\bar{\phi}$, to define the direction PQ relative to the axial system $\bar{X}$, $\bar{S}$, $\bar{V}$. The angle QPZ is $\bar{\rho}$ and $\bar{\phi}$ is the angle between two planes intersecting in the line PZ. (This is illustrated in Fig. 1.22 as well as Fig. 1.21.) The first of these two planes is that which contains the axes $\bar{S}$ and $\bar{V}$ and the second is a plane containing PQ and $\bar{V}$. This definition of $\bar{\rho}$ and $\bar{\phi}$ is analogous to that commonly used for ρ and ϕ. The definition of the orientation has been given here in a general form and applies to non-equatorial as well as equatorial rekhas. However, most of the diffuse reflection measurements are made around relps which lie in the equatorial plane. In this case $\bar{X}$ and $\bar{S}$ both lie in the equatorial plane. For a given value of $\bar{\rho}$, the line PQ must lie on the surface of a cone having P as its apex, $\bar{V}$ as its axis, and having a semi-angle of $\bar{\rho}$. This cone intersects the surface of the reflecting sphere in a curve AB (Figs. 1.21, 1.22), which is nearly an hyperbola. To each value of $\bar{\rho}$ there corresponds such an hyperbola. The line PI cuts the surface of the reflecting sphere in the point N and the plane containing axes $\bar{X}$ and $\bar{S}$ cuts the reflecting sphere in the arc NM. For a given value of $\bar{\rho}$, the greater the distance NP, the further is the hyperbola AB, corresponding to $\bar{\rho}$, from the arc NM. The ratio PN/NI is denoted s (regarded as positive when P lies outside the reflecting sphere), and is a quantity determining the scale of the diffuse reflection pattern. A chart has been constructed giving the hyperbolae for values of $\bar{\rho}$ ranging from 90° to 20°, for a value of s equal to $+0{\cdot}02$. This chart is shown in Fig. 1.23, and in Appendix 2 is given the method of computing the coordinates of the curves from which such charts can be drawn on

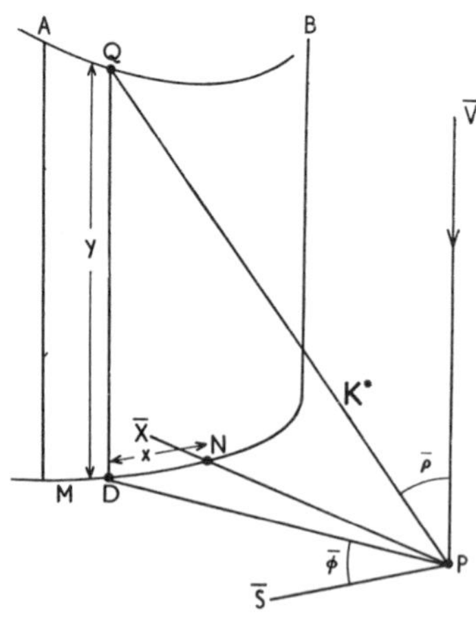

FIG. 1.22 Diagram showing the geometrical relation between the coordinates $\bar{\rho}$, $\bar{\phi}$ and x, y for a point Q. The relp P is lying on the equatorial plane.

any desired scale. The line NP in Fig. 1.23 gives the shortest distance from the relp to the reflecting sphere. Rekhas having a constant $\bar{\Phi}$ value lie in a single plane containing the axis $\bar{V}$ and this plane cuts the surface of the reflecting sphere in a curve, CQD, which intersects NM orthogonally (Figs. 1.21 and 1.22). The curves of constant $\bar{\Phi}$ are the almost vertical lines of Fig. 1.23.

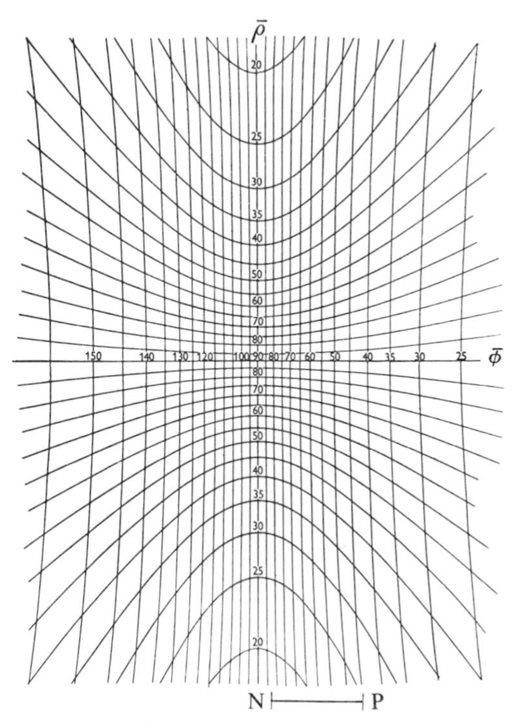

FIG. 1.23 $\bar{\rho}$-, $\bar{\Phi}$-chart constructed for $s = +0{\cdot}02$, as used in interpreting diffuse reflection photographs.

A third magnitude of great importance in the interpretation of diffuse reflection photographs is the distance PQ (Fig. 1.22), which is the same as K^* in the theoretical sections. The curves of constant K^* are circles about N on the reflecting sphere and almost circles on the equatorial photograph. Just as stereographic and gnomonic projections are usually made on a standard size of circle, namely, $2\frac{1}{2}$ in. or 5 cm diameter, so it is convenient to use a 5 cm scale for representing graphically the intersection of rekhas with the photographic film. The charts (Figs. 1.23, 1.24) of $\bar{\rho}$, $\bar{\Phi}$, and K^* are conveniently drawn on the basis of the distance $NP = 5$ cm. This implies that for a value of s equal to $+\ 0{\cdot}02$, the radius of the reflecting sphere is 250 cm.

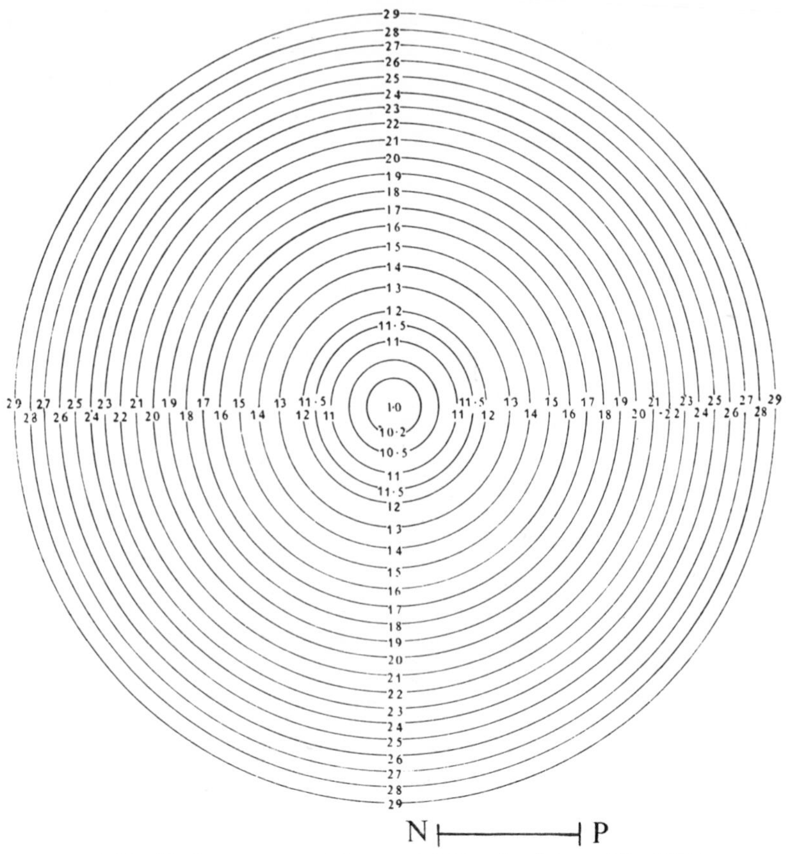

FIG. 1.24 K^*-chart constructed for $s = +0{\cdot}02$ and a distance NP of 10·0 cm.

When the value of s is non-standard the values of $\bar{\rho}$ and $\bar{\Phi}$ can be found from the cartesian coordinates x, y (Fig. 1.22) of any point in the diffuse spot. Neglecting the curvature of the reflecting sphere, we have

$$x = ND = NP \cot \bar{\Phi} = s \cot \bar{\Phi}, \tag{1.10}$$

$$y = QD = DP \cot \bar{\rho} = NP \operatorname{cosec} \bar{\Phi}.\cot \bar{\rho} = s.\operatorname{cosec} \bar{\Phi}.\cot \bar{\rho}. \tag{1.11}$$

From a knowledge of s and measurement of x and y the angles $\bar{\Phi}$ and $\bar{\rho}$ can be calculated, using equations (1.10) and (1.11). Further, we have

$$y = QD = QP \cos \bar{\rho} = K^* \cos \bar{\rho}, \tag{1.12}$$

and from this equation K^* may be found. For most purposes the approximation involved in neglecting the curvature of the reflecting sphere does not introduce any serious error. If, however, an exact calculation is required the methods described in Appendix II in connection with the construction of the standard charts may be used.

In order to use these charts it is necessary to make the value of $(i-\theta)$ correspond to the value of s for which the $\bar{\rho}$-, $\bar{\phi}$-, and K^*-charts have been worked out. The value of $(i-\theta)$ which corresponds to a particular value of s, when P lies in the equatorial plane, can be determined as follows. The letters P, N, I, O (Figs. 1.25, 1.26) have the same significance as in Fig. 1.22. We

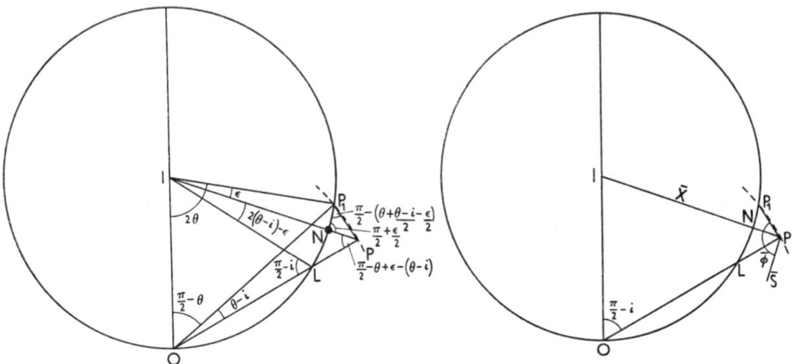

FIG. 1.25 Diagram illustrating the geometry of the diffuse pattern in relation to the relp and reflecting sphere.

FIG. 1.26 Diagram showing the angle $\bar{\phi}$ for the Laue spot L.

require to find PN in terms of $(i-\theta)$ or $(\theta-i)$. If $OP_1 = OP$, then angle $IOP_1 = \pi/2-\theta$ and angle $P_1OP = (\theta-i)$. We put angle $P_1IP = \varepsilon$. Then from the triangle NPP_1 we obtain

$$NP/PP_1 = \cos\left(\theta+\frac{\theta-i}{2}-\frac{\varepsilon}{2}\right)\Big/\cos\frac{\varepsilon}{2}.$$

Since ε seldom exceeds 3° we may take $\cos \varepsilon/2$ as equal to unity. Further, $PP_1 = 2\sin\theta.(\theta-i)$ and hence

$$s = NP = 2\sin\theta.(\theta-i)\cos\left(\theta+\frac{\theta-i}{2}-\frac{\varepsilon}{2}\right) \tag{1.13}$$

or, approximately,

$$s = (\theta-i).\sin 2\theta. \tag{1.14}$$

Similarly,

$$NP_1/PP_1 = \sin\left(\theta+\frac{\theta-i}{2}-\varepsilon\right)\Big/\cos\frac{\varepsilon}{2},$$

and hence

$$\varepsilon = NP_1 = 2.\sin\theta.(\theta-i).\sin\left(\theta+\frac{\theta-i}{2}-\varepsilon\right) \tag{1.15}$$

or, approximately,

$$\varepsilon = 2\sin^2\theta.(\theta-i). \tag{1.16}$$

If the expression (1.14) does not give a sufficiently accurate value of s, the value of ε obtained from (1.16) may be used in expression (1.13). The sign of s is taken as positive when $\theta > i$ and the point P lies outside the reflecting sphere. Thus the angle through which the crystal must be rotated away from the Bragg setting to give a particular value of s can be easily calculated.

A check on the accuracy of the setting of the crystal is afforded by the separation on the photograph of the Bragg and Laue spots. In Fig. 1.26 the Bragg spot occurs in a position on the film corresponding to P_1. (We suppose the film to coincide with the surface of the reflecting sphere.) The Laue spot occurs during the taking of the diffuse reflection photograph in the position denoted by the point L, which is at the point where the rel-vector OP passes through the surface of the reflecting sphere. The distance NL (Fig. 1.25) is equal to $[2(\theta - i) - \varepsilon]$ and, from the radius of the camera, r, the distance from the Laue spot to the point N can be determined as $r.[2(\theta - i) - \varepsilon]$. Owing to the fact that the Laue spot is in general produced by radiations that are much more penetrating than the characteristic radiation, the Laue spot may be spread out to a greater or lesser extent depending on the absorption of the crystal. The centre of the Laue spot may, therefore, be a less reliable guide to the position of the point N than a Bragg spot at the position P_1. To avoid interfering with the diffuse pattern the Bragg reflection can be impressed on another film, or on a neighbouring portion of the same film. A modified Weissenberg goniometer has been used for this purpose (Hoerni, 1952).

1:3.1.5 *Use of the $\bar{\rho}$-, $\bar{\Phi}$-, K^*-charts in interpretation*

The diffuse reflection photograph must first be studied by means of a microdensitometer in order to determine the variation of intensity of reflection over the whole area of the diffuse spot. From this investigation a contoured density map can be plotted on a scale corresponding to the charts in use, for example, by making the distance of the relp from the surface of the reflecting sphere 5·0 cm. The charts can be superposed on such a contoured map and the values of $\bar{\rho}$, $\bar{\Phi}$, and K^* at any point of the diagram determined. From the known orientation of the crystal during the taking of the photograph, the direction of the rekhas corresponding to particular values of $\bar{\rho}$ and $\bar{\Phi}$ can be found. For this purpose a stereogram may be employed or the axes may be transformed by the usual set of direction cosines. The $\bar{\rho}$-, $\bar{\Phi}$-values serve to define the point on the photograph corresponding to the intersection of a [100], [110] or [111] rekha with the reflecting sphere. The K^*-chart enables us to read off the length of the wave vector corresponding to any point. Several rekhas are usually studied and their relative densities, or X-ray intensities, compared.

1:3.1.6 *Divergence corrections*

When the diffuse scattering points form a cloud of slowly varying density round the relp, the intensity at any point of the diffuse spot is a correct measure of the scattering power of the corresponding element of volume in reciprocal space. When, however, the diffuse scattering corresponds to a disk, a rod, or a point in reciprocal space, special consideration must be given to the interpretation of the observed density at any point of the pattern (Hoerni and Wooster, 1953). We shall not pursue this matter further here as the examples chosen in Chapter II refer only to the cloud-like reciprocal distributions. A divergence correction of another kind is dealt with in section 2:4.1. This is the effect of an incident beam, which is not infinitely narrow or strictly parallel, in blurring the pattern.

1:3.2 *Ionization method*

The apparatus used for studying the intensity of the diffuse reflection may conveniently be divided into a mechanical part and an electronic part. The mechanical part is concerned with the setting of the crystal and the setting of the detector, which may be a Geiger counter, proportional counter, scintillation counter, or ionization chamber, and the electronic part is concerned with the recording of the intensity of the reflected X-rays.

1:3.2.1 *Mechanical part of the diffractometer*

A number of diffractometers have been made and accounts of them have been published. The one described here† has been used by the author (Wooster 1960*a*); it contains most of the features of other diffractometers and will serve to indicate the essential principles of such instruments.

It consists of a main base, *A* (Fig. 1.27), containing two large worm-wheels, *B*, each having 360 teeth, and two worms bearing on these wheels provided with graduated drums, *C*. The rotation of each worm through 360° corresponds to the rotation of the worm-wheel through 1°. The division of the graduated drum into 60 or 100 parts provides for the reading of the angle in minutes or hundredths of a degree, and counters, *D*, driven by the worm-bearing on the large wheel, indicate the rotation of the worm-wheel in degrees and tenths of degrees. The two worm-wheels, *B*, may be connected together by gearing so that the upper one rotates at twice the angular speed of the lower one. In order to avoid any backlash the worms are spring-loaded on the worm-wheels. This is achieved in the following way: on the left-hand side, the shaft of each worm has a sphere turned on it, and this is held between two adjustable cones which are tightened up so that they hold, but do not grip, the sphere. On the right-hand side, a shaft carrying the worm moves freely in a parallel-sided slot and is pressed by a spring towards the worm-wheel. On the left-hand side, each worm is driven through a universal joint which does not exert any influence on the position of the worm relative to the worm-wheel. In this way a constant pressure can be maintained between the worm and the worm-wheel under all circumstances. The large worm-wheels are supported

† Made by Crystal Structures Ltd., Cambridge.

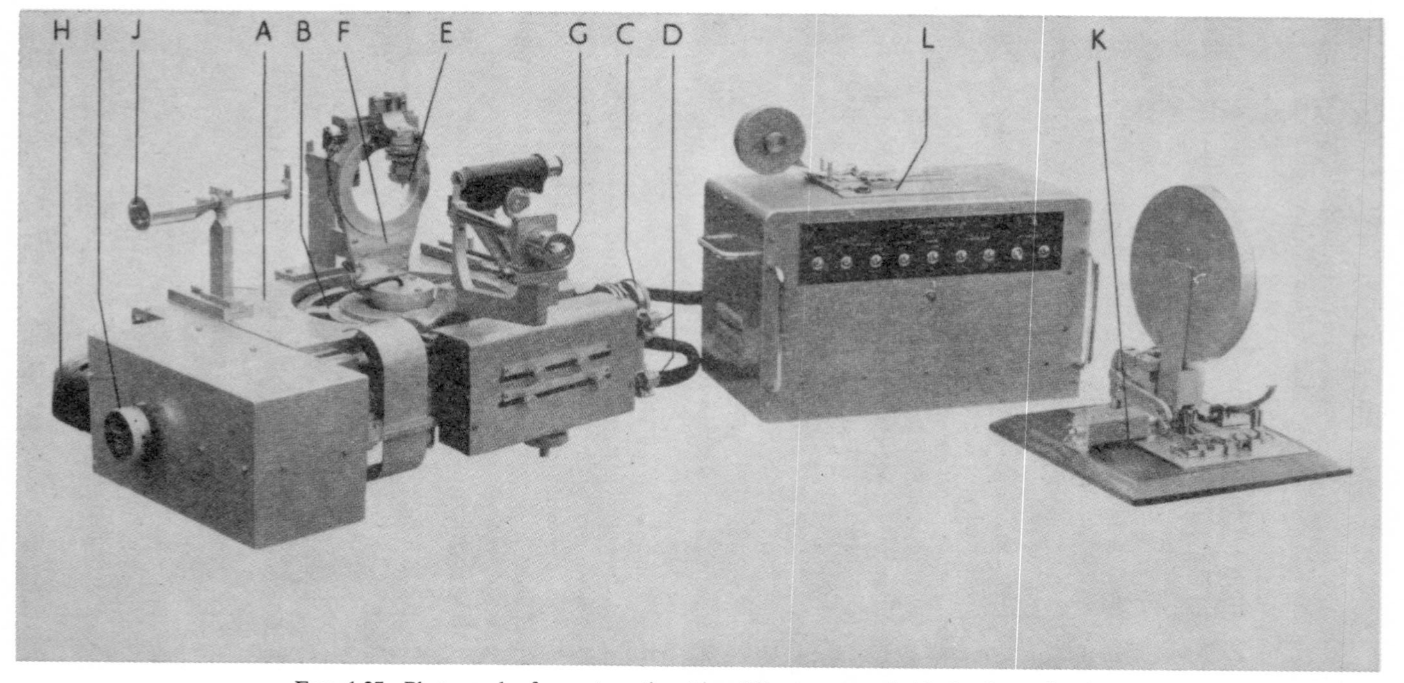

FIG. 1.27 Photograph of an automatic setting diffractometer suitable for the study of a large number of points in reciprocal space.

on cone bearings, one cone within the other; the lower wheel rotates the crystal, *E*, which is carried on a goniometer head mounted on the vertical circle, *F*. The upper wheel rotates the detector, *G*. The drive for these wheels is provided by a 40-watt synchronous electric motor, *H*, which drives various gear-wheels so as to provide the speeds required for the particular investigation in hand. The speed of rotation of the motor is 3,000 r.p.m. and this is reduced by a worm gear to 100 r.p.m. This speed is transmitted by one train of gear-wheels directly to the instrument so that the detector can be rotated at 100 r.p.m. when required. A second train of gear-wheels coming from the motor passes through a 6-stage gearbox, *I*, by means of which a driving speed for the instrument of 8, 4, 2, 1, $\frac{1}{2}$, $\frac{1}{4}$, degrees per minute can be supplied. The changeover of one of these lower speeds to the faster speed of 100 r.p.m. can be achieved by the operation of an electrically controlled dogclutch. The fast speed is used for moving the crystal or the detector rapidly from one position to another. The slow speeds are used for the careful survey through a small angular range. Many observations on diffuse reflection have been carried out, not by continuous rotation of the detector or the crystal, but by setting the crystal and the detector in particular positions and making observations over a certain given time. When a large number of settings of the crystal and the detector have to be made one after another, it is convenient to use the automatic setting and registering facilities of the instrument illustrated in Fig. 1.27. The setting of the crystal and the detector is effected by using a punched paper tape. This is capable of having any combination of five holes punched in it and any desired programme may be arranged.

Instructions to perform any combination of these movements may be transferred to the paper tape by the punch *K*. A tape reader, *L*, and a series of relays beneath it operate the clutches which control the movements of the various gear-wheels. The second Geiger counter, *J*, is used to monitor the intensity of the incident X-ray beam.

1:3.2.2 *Electronic part of the diffractometer*

The choice of the detector determines the kind of electronics required. In the early forms of the instrument an ionization chamber was employed. The current available for this was always small and the measurements of this small current gave rise to considerable experimental difficulties and relatively low accuracy. Later development introduced the Geiger counter, which registered almost every X-ray quantum absorbed in the counter chamber. This was much more sensitive but involved the development of the recording of individual pulses. For this purpose two principal devices are employed, one a ratemeter the other a scaler. The ratemeter measures the average rate of arrival of the pulses and the scaler gives the total number of pulses counted within a given time. The ratemeter indicates the charge accumulated on a condenser due to the arrival of the pulses and this charge fluctuates according to the rate of arrival of the pulses and the time constant of the circuit. If a large time constant is used then the meter indicating the potential of the tank condenser remains relatively steady, but on the other hand it does not follow changes in the intensity of the X-rays due to the movement of the crystal when these are rapid compared with the time constant of the electrical system. If the time constant is made small then the fluctuations of the needle for a given rate of arrival of the pulses is correspondingly rapid. In practice therefore, the rate at which the crystal can be moved has to be such that when the appropriate time constant is chosen the needle remains reasonably steady. An automatic record can be made on one of the pen-recorders when a ratemeter is employed, and this is the common arrangement for many modern diffractometers. For quantitative observations of diffuse reflections it is often more convenient to use a scaler. This gives an actual number of counts and the results can be more readily worked out from such a number than from the kind of trace which is obtained with a pen-recorder. It is always necessary to make allowance for the background count, which may be due to cosmic rays, Compton scattering, or various other causes associated with the crystal or the auxiliary apparatus. When a scaler is used this background can be readily determined simply by preventing the radiation coming from the crystal reaching the counter but otherwise leaving everything else unchanged. The readings obtained from the scaler can also be automatically recorded provided that a print-out unit is available which will convert the reading in the scaler to numbers on a sheet of paper. One of the important questions connected with the recording of X-ray intensities is the choice of the detector.

The Geiger counter, although as sensitive as any detector can be since it registers practically all the pulses which are incident on it, has nevertheless certain inherent disadvantages. Each pulse lasts for a time measured in hundreds of micro-seconds and during this time the counter is unable to register the entrance of any further X-ray quanta. The so-called 'dead time' sets a limit to the rate at which the Geiger tube can count, and also introduces an inaccuracy which varies with the rate of counting. This difficulty is overcome by making use of a proportional counter which is constructed in a way very similar to that of the Geiger tube itself but operated at a potential which is in a different region of the characteristic curve. Whereas the number of counts registered by the Geiger counter is practically independent of the applied voltage over a range of about a hundred volts the proportional counter is operated below the threshold value characteristic of the Geiger tube and the current through it is highly dependent on the voltage applied. However, provided this voltage is kept constant this is of no importance and the dead time associated with the proportional counter is very much less than that of the Geiger counter, being of the order of a micro-second. The rate of counting which is possible with a proportional counter is therefore much higher than is possible with a Geiger tube, and moreover, no correction need in general be made for the dead time of the tube. Similar advantages are to be found with the scintillating counter. A scintillating counter is made up of a crystal which need be no more than a few millimetres in each of its linear dimensions and this is coupled to a photo-multiplier which registers the light pulses generated by the impact of X-ray quanta. The disadvantages of the scintillating counter are due to the fact that the thermal excitation of light pulses is comparable in intensity with the excitation due to the quanta of the softer X-radiations. For harder and more energetic X-ray quanta, such as those from molybdenum and silver, this difficulty does not arise and the scintillating counter is likely to prove very useful.

The advantage of both the proportional counter and the scintillation counter is the dependence of the magnitude of the electrical pulse on the energy of the absorbed X-ray quantum. This can be utilized in order to make the counter partially selective to one radiation rather than another. For instance, if copper radiation is used to study an iron-containing material the fluorescent iron radiation is excited. With a Geiger counter it is impossible to distinguish between the entry of X-ray quanta due to the fluorescent radiation from the iron and the X-ray quanta due to the reflection of copper radiation from the crystal. Other means have to be adopted in order to avoid the effect due to the fluorescent radiation. With a proportional or scintillating counter, however, the energy of the quanta determines the magnitude of the voltage peak which is generated due to the X-ray quantum and, by using a discriminator which transmits to the rest of the indicating equipment only pulses of magnitude greater than the predetermined value, the effects of the fluorescent radiation can to a large extent be eliminated, without impairing the sensitivity of the apparatus for recording the characteristic copper radiation.

1:3.2.3 *Stabilization of X-rays*

One of the great difficulties about the diffractometer method is that the observation tends to be very dependent upon maintaining a constant intensity of incident X-rays. This is in striking contrast to photographic methods in which every part of the pattern receives a constant fraction of the incident intensity whatever that may be. Stabilization of the X-ray beam can be effected in a number of ways. Some of the commercial X-ray generators are provided with electronically stabilized X-ray outputs, though it is always necessary to be very careful that the criterion according to which the beam is adjusted is one which is relevant for the experiment in hand. For instance, even if the intensity of the X-ray beam emerging from a particular window is constant, a slight movement of the collimating slit relative to the focus of the X-ray tube may result in a considerable change in intensity. The same objection can be raised to any method of stabilization which depends on controlling the voltage and filament current of the X-ray tube. However, it is often valuable to have an X-ray beam which is constant in intensity and it is not difficult, in general, to maintain the relative position of the diffractometer and the X-ray tube constant. The stabilization of the output of an X-ray tube can be effected by valve control. It can also be attained by operating the high tension transformer from a local alternating current generator which is itself completely independent of the mains supply. For this pupose it should preferably be driven by a direct current motor supplied from a battery.

An alternative to stabilization of the X-ray beam is the use of a monitor. A monitor contains a second Geiger tube or proportional counter and usually receives a portion of the direct beam, either by (Bragg) scattering from a piece of metal foil, through which the whole of the incident beam passes, or by scattering from a crystal set at a particular angle. The quantity measured in this case is the ratio of the intensity of the diffusely reflected beam to the intensity of the beam scattered from the aluminium foil and registered by the monitoring tube. A convenient system embodying this principle can be made using two scalers. The counts from the monitor are registered in one scaler and when a fixed number of counts is reached, say ten thousand, the operation of both scalers is stopped. The reading of the scaler corresponding to the diffuse reflection is in this case independent of the intensity of the incident X-ray beam. The variation in the intensity of the incident beam simply results in a change in the time of counting the predetermined number of pulses.

1:3.2.4 *i*-, *ϕ-charts for setting crystals*†

A photograph of diffuse X-ray reflection provides a survey over a considerable area of reciprocal space. A diffractometer records the diffuse intensity arising from a small area of the reflecting sphere, or, if the divergence of the X-ray beam is taken into account, from a small volume element of reciprocal space. The first problem is, therefore, to determine which volume element is being studied when the crystal and the detector are placed in any given settings. In Fig. 1.28 there is shown the usual diagram of the reflecting circle, *OP*, and relp, *P*. If the crystal remains stationary diffuse reflections arise in directions such as *IQ*, defined by joining the centre of the reflecting sphere to the points on the surface near to the relp *P*.

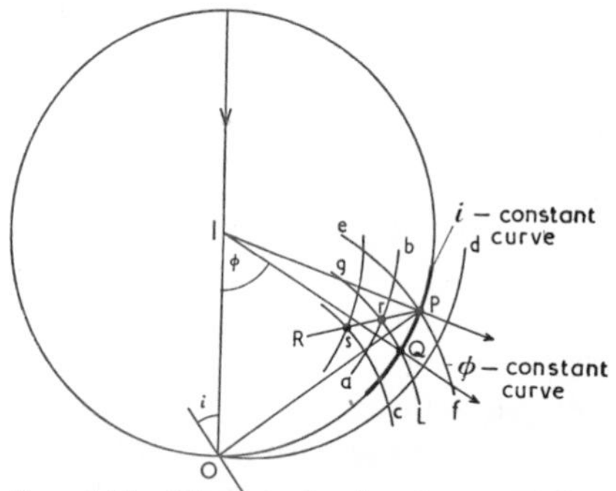

FIG. 1.28 Diagram showing the generation of constant-*i* curves from the successive positions of the reflecting sphere and of constant-ϕ curves from the rotation of the relp about the origin.

† It should be noted that there is no connection between the symbols ϕ and $\bar{\phi}$. The former refers to the deviation of the X-ray beam and the latter to the direction of a rekha.

The circle OQP corresponds to diffuse reflections for a given value of the angle of incidence, and is called a *constant-i curve*. If the crystal is rotated from this position to a slightly different one then the line OP rotates relative to the diameter, IO, of the reflecting circle. Although in reality it is the reflecting circle which remains stationary and the rel-vector which rotates it is more convenient in making geometrical constructions to suppose that the reciprocal lattice remains stationary and that the reflecting circle rotates about O. Adopting this convention we can say that the reflecting circle Ocd corresponds to the setting of the crystal after a certain small rotation. For each value of the angle i there is a corresponding arc cd. A chart can be constructed on a radius IP = 50 cm and with constant-i curves drawn at 1° intervals.

We shall next suppose the detector to be maintained in a fixed orientation relative to the incident beam and the crystal to be rotated over a small angular range. For the diffusely reflecting point Q, the X-rays are deviated through a fixed angle, $OIQ = \phi$, and as the angle i is changed the points which reflect at this angle ϕ lie on a circle concentric with O containing the arc gL. Thus to each value of ϕ there corresponds a circular arc such as gL. If the radius IP is 50 cm then the radius $OP = 100 \sin \phi/2$ cm. Constant-ϕ curves may be drawn corresponding to intervals of 1° on the same diagram as the constant-i

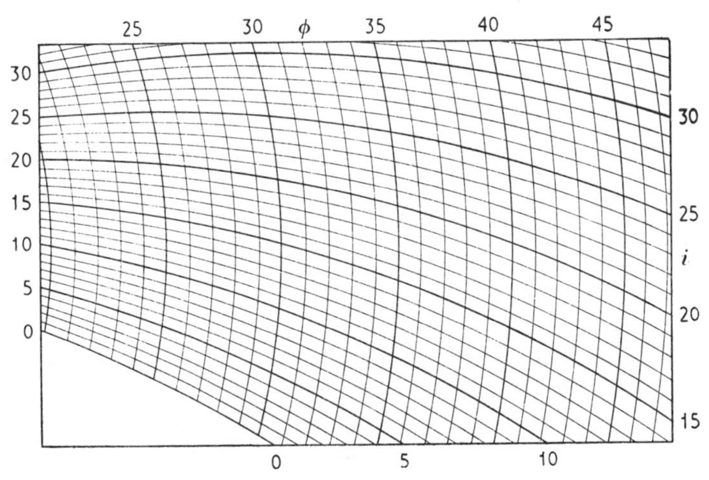

FIG. 1.29 i-, ϕ-chart for the range in ϕ of 20°–47°. Radius of ϕ-circle $= 2IP.\sin \phi/2$ (Fig. 1.28).

curves. The combination forms the so-called i-ϕ-chart, one of which is given in Fig. 1.29 and two others in Appendix I. These charts cover the ranges in ϕ, 20°–47°, 42°–70°, and 66°–100°. Such charts are used in preparing a programme of measurements of diffuse reflections. Suppose, for example, that the diffuse intensities at several points r, s, along the rekha PR (Fig. 1.28) are required.

The values of i and ϕ for the curves which pass through the point r are read off from the chart and the crystal and detector are set to corresponding positions. After the intensity of reflection has been measured, the settings are changed to correspond to the point s, and the observation of intensity repeated. Thus the variation of the intensity of reflection along any rekha passing through a given relp P can be investigated. It is sometimes necessary to follow a curve surrounding the point P corresponding to an iso-diffusion surface, i.e. the surface for which all points give an equally strong diffuse reflection. The same chart can be used for this purpose by drawing the curve round P and noting the i-ϕ-values which correspond to the successive points round the iso-diffusion surface.

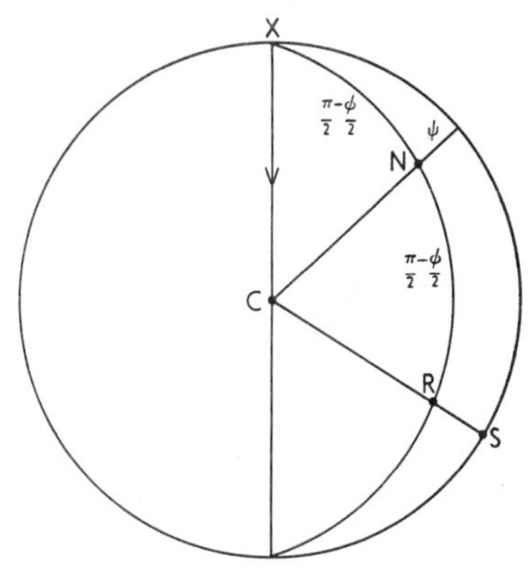

FIG. 1.30 Diagram showing the geometry of a non-equatorial reflection.

The above discussion relates to the equatorial plane. Most diffractometer observations are made in the equatorial plane but this is not essential. Provided the diffractometer is fitted with a detector which can be rotated round an arc in a vertical plane then reflections which do not lie in the equatorial plane can be recorded. The angle between the rel-vector concerned in the reflection and the equatorial plane is denoted ψ. In Fig. 1.30 is shown a stereogram giving X, the direction of the incident beam, N, the normal to the reflecting planes tilted at an angle ψ with respect to the equator, and R, the direction of the reflected X-rays. It may be shown that

$$\sin RS = 2 \sin \frac{\phi}{2} \sin \psi,$$

and the detector must be tilted through an angle RS relative to the equator.

An alternative method of orientating the crystal makes it possible to maintain the detector always in the equatorial plane. The crystal is mounted on a goniometer head which can be rotated round a vertical circle, i.e. about a horizontal axis (see Fig. 1.27). Any rel-vector may be brought into the equatorial plane by using this vertical circle and in this case it is not necessary to raise the detector above the equatorial plane.

1:3.2.5 *Charts for skew corrections*

A flat plate of crystal is commonly used in diffuse reflection measurements. This may be a natural face or an artificially prepared face cut parallel to a particular lattice plane. At the Bragg angle of incidence $i = \theta = \phi/2$, but in general, during a survey of diffuse reflections $i \neq \theta/2$. The inequality of the angles of incidence and reflection gives rise to an important change in the intensity of reflection. This may be seen as follows. A ray PQR (Fig. 1.31) is

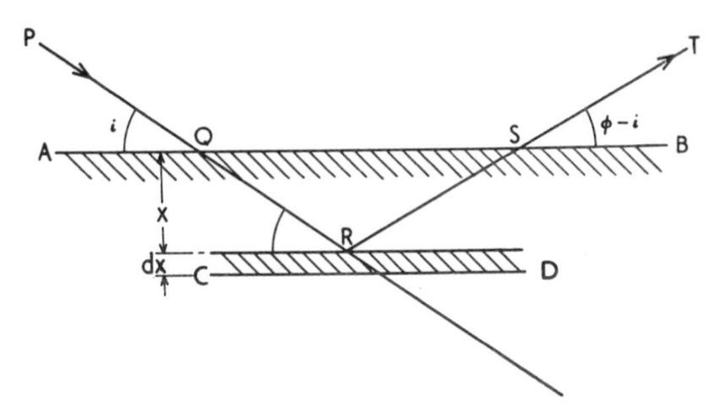

FIG. 1.31 Diagram showing skew reflection from an element of volume CD under the reflecting surface AB.

incident on the crystal surface AB at an angle i and is reflected from a layer CD parallel to the surface and situated at a depth x below the surface. The thickness of the layer CD is dx. If the intensity of the incident rays per unit area of cross section is I_0 and μ is the linear absorption coefficient for X-rays, then the intensity reaching the point R is I_R, where

$$I_R = I_0 e^{-\mu QR}.$$

The volume of the layer CD irradiated by a beam of unit area is equal to $dx/\sin i$ and the reflected intensity I_r leaving this thin layer in the direction RS is

$$I_r = I_0 e^{-\mu QR} \frac{dx}{\sin i} . P,$$

where P is the reflecting power per unit volume of the material under these conditions. To reach the surface these reflected rays must pass through a thickness RS and hence the intensity on emergence, I_s, is given by

$$I_s = I_0 e^{-\mu QR} \frac{dx}{\sin i} . P . e^{-\mu RS}$$

$$= \frac{I_0}{\sin i} . P . e^{-\mu(QR+RS)}\, dx.$$

Now $QR = x/\sin i$ and $RS = x/\sin(\phi - i)$.

Hence

$$I_s = \frac{I_0 P}{\sin i} \exp\left[-\mu x\left\{\frac{1}{\sin i} + \frac{1}{\sin(\phi - i)}\right\}\right] dx.$$

The total intensity I_t due to contributions from all such layers as CD is given by

$$I_t = \int_0^\infty I_s\,dx$$

$$= \frac{I_0 P}{\sin i}\cdot\frac{1}{\mu\left\{\dfrac{1}{\sin i}+\dfrac{1}{\sin(\phi-i)}\right\}}$$

$$= \frac{I_0 P}{\mu}\cdot\frac{1}{\left\{1+\dfrac{\sin i}{\sin(\phi-i)}\right\}}.$$

The value of I_t when $i = \phi/2$ is put equal to I_d and then

$$I_d = \frac{I_0 P}{\mu}\cdot\frac{1}{2}$$

$$= I_t.\frac{1}{2}\left\{1+\frac{\sin i}{\sin(\phi-i)}\right\}. \tag{1.17}$$

The factor $\frac{1}{2}\{(1+\sin i)/\sin(\phi-i)\}$ is called the skew correction. The observed intensity I_t must be multiplied by the skew correction to obtain I_d, the intensity which would be observed if the angles of incidence and reflection were equal. It will be seen that when $i > \phi/2$ the skew factor is greater than unity and when $i < \phi/2$ it is less than unity. When $i > \phi/2$ the volume of the crystal irradiated is less than when $i = \phi/2$. For both these reasons I_t is less than I_d. Charts have been constructed giving the magnitude of this correction for all values of i and ϕ. These charts have been drawn on the same scale as the $i-\phi$-charts so that they can be used on the same diagrams of the reciprocal lattice. An example of such a skew correction chart is given in Fig. 1.32 and further examples for the same ranges of angles as apply to the $i-\phi$-charts are given in Appendix I.

1:3.2.6 *Divergence corrections*

The finite size of the slits between the crystal and the X-ray tube and also between the crystal and the detector permits X-rays having a certain divergence to enter the detector and be counted together. The experimental observation does not, therefore, correspond to the intensity of diffuse scattering from a point in reciprocal space but it corresponds to the effect integrated over a certain volume of reciprocal space. The greater the divergence of the X-ray beam the greater the volume of reciprocal space over which the integration is effected. The divergences may be divided into three components denoted respectively i-, ϕ-, and ψ-divergence. The i-divergence is the maximum angle between the rays falling on the crystal when they are projected on a horizontal plane. The ϕ-divergence is the maximum angle between the reflected rays, also projected on a horizontal plane, which can enter the detector at any given setting. The ψ-divergence is the maximum angle between the reflected rays,

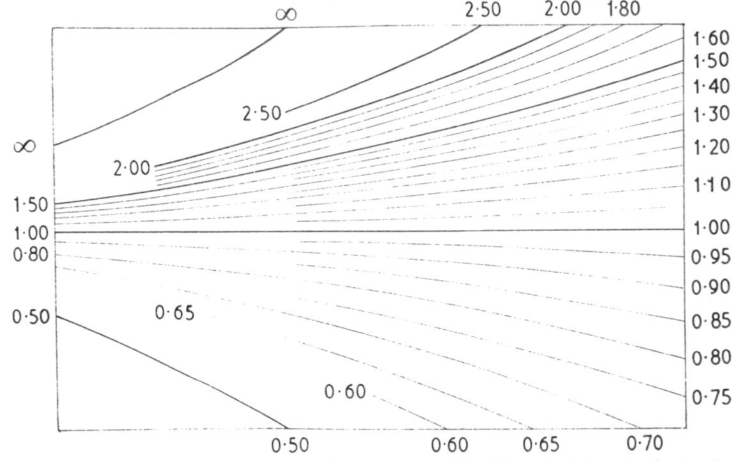

FIG. 1.32 A skew correction chart for the range 20°–47° in ϕ, giving the factor by which the observed intensity must be multiplied to correct for the skewness of the reflection.

projected on to a vertical plane passing through the crystal and the detector, which can be recorded at the given setting. The magnitude of divergence will vary from one experiment to another but generally for the angle i it is about 30′, for the angle ϕ about 1°, and for the angle ψ about 5°.

The determination of the correction for i-divergence

We shall first consider the method of correcting for the i-divergence. Keeping the ϕ- and ψ-divergencies constant, i.e. keeping the same size of aperture in front of the detector, we plot a curve, AB (Fig. 1.33*a*), showing the variation of the diffuse intensity, $D(i)$, with the setting on the diffractometer corresponding to the angle i. The range of i over which this curve is to be plotted is usually not more than 1° on either side of the given value of i. The crystal is now rotated to the Bragg setting and the detector is set to receive this reflection. While the detector is kept in a fixed setting, the crystal is set in a number of positions on either side of that giving the maximum intensity, and a curve (Fig. 1.33*b*) is plotted giving the variations of the intensity, $B(i)$, with the angle i read on the diffractometer. As will be seen from Figs. 1.33*a* and 1.33*b*, the diffuse intensity varies slowly with i whereas the Bragg reflection varies rapidly with i in the neighbourhood of the Bragg setting. Each point on the curve AB can be regarded as being produced by many planes slightly inclined to the reciprocal lattice planes which give the Bragg reflection. At a point A,

Fig. 1.33*a*, one lattice plane is in the Bragg setting and its contribution to the intensity is represented by the small copy *CD* of the Bragg intensity curve drawn at the same angle *i* as that of point *A*. This particular plane contributes the maximum possible diffuse intensity at point *A*. Other planes, nearly parallel to the *A*-set, contribute an amount corresponding to the departure of

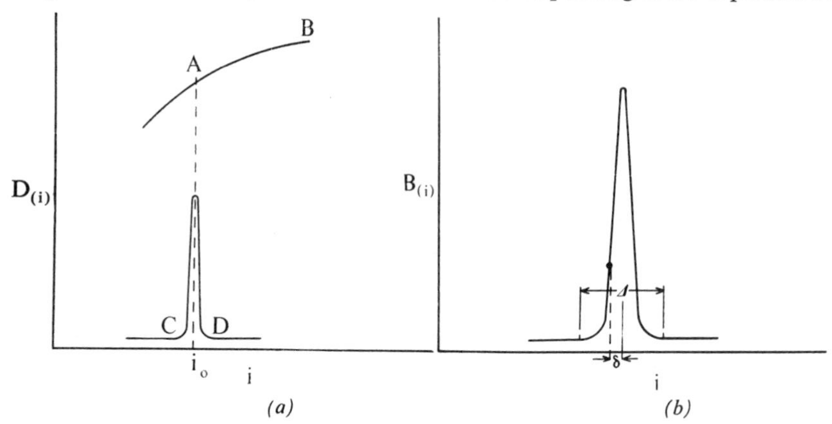

FIG. 1.33*a* The curve *AB* corresponds to the variation of intensity, *D*, across the diffuse reflection. The curve *CD* gives the variation of intensity across the Bragg reflection.

FIG. 1.33*b* The curve corresponds to the variation of intensity across a Bragg reflection. The curve *CD* is a copy, reduced to an appropriate scale, of the Bragg reflection curve.

the angle *i* from that value which corresponds to their maximal contributions. Thus at any point *A* the value of $D(i)$ is the sum of contributions, each one of which is the product of the maximum intensity of diffuse scattering from a lattice plane, and an ordinate of the curve *CD* (Fig. 1.33*a*). Let us denote by $D(i)$ the true intensity curve of the diffuse reflection due to the disturbed lattice without any modification due to the shape of the $B(i)$ curve. The angular breadth of the $B(i)$ curve is denoted by Δ, and the angular separation of any point on the $B(i)$ curve from the central maximum is called δ (Fig. 1.33*b*). The particular value of *i* at the point *A* on curve $D(i)$ is denoted by i_0. The elastic waves corresponding to a Bragg setting $(i_0+\delta)$ give a contribution at *A* equal to

$$D'(i_0+\delta)B(\delta),$$

where $B(\delta)$ is the ordinate of the $B(i)$ curve at a distance δ from the maximum. The total effect is thus the integration of all these contributions from the separate elastic wave trains, namely,

$$D(i_0)=\int_{-\Delta/2}^{+\Delta/2} D'(i_0+\delta)B(\delta)\,d\delta. \qquad (1.18)$$

It is here assumed that the Bragg reflection curve is symmetrical about its centre point. This assumption is usually justified. If the curve $D(i)$ plotted in Fig. 1.33*a* is a straight line then the integration given by equation (1.18) gives a value of $D(i_0)$ which is proportional to $D'(i_0)$. If, however, the curve $D(i)$ and therefore the curve $D'(i)$ is not straight (Fig. 1.33*a*), the contributions at points $+\delta$ and $-\delta$ from the centre are not equal and their mean value is less or greater than that corresponding to the angle i_0. In Fig. 1.34 the

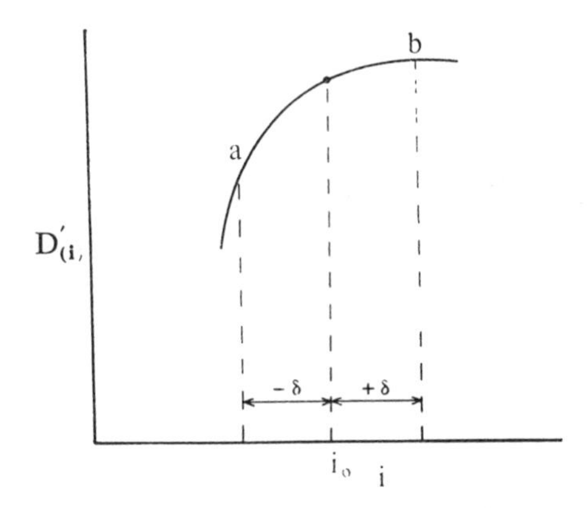

FIG. 1.34 Diagram showing the meaning of the quantity δ used in divergence corrections.

curvature of the $D'(i)$-curve results in a mean of the values, $D'(i_a)$, $D'(i_b)$ of $D'(i)$ at points *a* and *b* respectively, which is less than $D'(i_0)$. Considering only the contributions from points *a* and *b*, Fig. 1.34, the error in taking $D(i)$ as proportional to $D'(i)$ expressed as a fraction of $D(i_0)$ is

$$\{\tfrac{1}{2}(D'(i_a)+D'(i_b))-D'(i_0)\}/D'(i_0)$$

In those cases where the divergence correction is small we may replace the primed letter D by the unprimed D. From an experimental curve such as that of Fig. 1.33*a*, which gives $D(i)$ as a function of i, we may calculate the quantity $G(\delta)$ for each value of δ between 0 and $\pm\Delta/2$ according to the expression

$$G(\delta) = \{\tfrac{1}{2}(D(i_0+\delta)+D(i_0-\delta))-D(i_0)\}/D(i_0).$$

The expression $G(\delta)$ gives the correction factor at every value of δ which must be applied to the experimental curve. To take account of the shape of the $B(i)$ curve we multiply each value of $G(\delta)$ by the corresponding $B(\delta)$ and integrate over the range $\pm\Delta/2$. To normalize the expression we divide by the integral of $B(\delta)$ over the same range and thus obtain the final correction term $Z(i_0)$ for the i-divergence, namely,

$$Z(i_0) = \int_{-\Delta/2}^{+\Delta/2} G(\delta)B(\delta)\, d(\delta) \bigg/ \int_{-\Delta/2}^{+\Delta/2} B(\delta)\, d\delta. \tag{1.19}$$

The evaluation of this correction factor is usually carried out graphically. The observed diffuse intensity at any given i_0 must be multiplied by a factor $(1+Z(i_0))$, which is greater or less than unity when the $D(i)$ curve is concave or convex respectively seen from below.

The correction for ϕ-divergence

A similar procedure may be used in order to find the correction necessary on account of the finite angular divergence $d\phi$ of the reflected beam which can enter the aperture in front of the detector. The crystal is set to reflect the Bragg maximum and then, with the angles i and ψ kept constant, the detector is turned through the angle Δ over which the Bragg reflection can be received. This angle is much greater than the corresponding Δ in the correction for i-divergence since the aperture in front of the detector will generally subtend an angle of 1° in the horizontal plane. From these observations the function $B(\delta)$ is obtained. The crystal and detector are now placed in the chosen settings for the diffuse observation and the variation of the intensity with change in ϕ is noted as the detector is moved through an angle rather greater than $\pm\Delta/2$. From the curve of the variation of the intensity with angular setting of the detector, the function $G(\delta)$ may be obtained. The correction factor can now be evaluated as for the i-divergence.

The correction for ψ-divergence

The crystal is set in the Bragg reflection position and, with the angles i and ϕ kept constant, the vertical circle is rotated so as to tilt the normal to the reflecting plane out of the horizontal plane. The variation of the intensity of the Bragg reflection with change in the angle ψ is observed and a $B(\delta)$ curve is plotted. The crystal and detector are now returned to the setting at which the diffuse intensity of reflection is to be measured. Still with i and ϕ constant, the intensity of diffuse reflection is measured as a function of ψ over a range greater than $\pm\Delta/2$ on either side of the chosen settings. From these observations a $G(\delta)$ is plotted and the correction factor is determined as before.

Finally all three correction terms can be added together to give a final correction factor.

1:3.2.7 *Absolute measurements*

To make full use of quantitative measurements of diffuse reflections it is often necessary to find $I(\mathbf{H}^*)$ in terms of the scattering from a single classical electron placed at the same point within each unit cell. A monochromatizing crystal is often used in this work and the X-rays reflected from it are partially polarized. If p_1 and p_2 are the amplitude components normal and parallel respectively to the plane of incidence, then, for unit intensity of the X-ray beam reflected from the crystal, at an angle θ,

$$p_1^2+p_2^2 = 1$$

and

$$p_1/p_2 = 1/\cos 2\theta,$$

or

$$p_1^2 = 1/(1+\cos^2 2\theta), \qquad p_2^2 = \cos^2 2\theta/(1+\cos^2 2\theta).$$

If a monochromatizing crystal is not used, so that the radiation is unpolarized, then

$$p_1 = p_2 = 1/\sqrt{2}.$$

The amplitude of scattering, ε, by a classical electron is given by the well-known formula

$$\varepsilon = \frac{e^2}{mc^2}\sqrt{(p_1^2+p_2^2\cos^2\phi)}, \tag{1.20}$$

where e and m are the charge and mass of an electron, c the velocity of light, and ϕ the angle through which the X-rays have been deviated by the scattering. Remembering the definition of first and second order thermal diffuse scattering powers (sections 1:2.2.3, 1:2.4), we see that the intensity diffusely reflected per unit cell per unit solid angle per unit incident intensity is $\varepsilon^2(D_1+D_2)$. (We can usually neglect any higher order of diffusely scattered radiation than the second.) The diffusely scattered intensity per unit solid angle per unit volume is thus

$$\varepsilon^2(D_1+D_2)/\tau,$$

where τ is the volume of the unit cell. We have seen (section 1:3.2.5) that when X-rays are diffusely reflected in the symmetrical setting, for which $\phi = 2i$ the ratio of the intensity of the diffusely reflected radiation, I_d, to that of the incident radiation, I_0, is given by

$$I_d/I_0 = P/2\mu,$$

where P is the fraction of the incident radiation scattered per unit volume by any small element of volume. In fact we have

$$P = \varepsilon^2(D_1+D_2)/\tau.$$

Under experimental conditions we are not dealing with parallel beams but the detector admits a certain solid angle, Ω, of the diffusely reflected radiation. Thus, finally, we have for the intensity of the diffusely reflected radiation

$$\frac{I_d}{I_0} = \varepsilon^2\frac{(D_1+D_2)\Omega}{2\mu\tau}. \tag{1.21}$$

The ratio I_d/I_0 may be given as the ratio of the number of counts registered per minute when the diffuse radiation is entering the detector to the number entering when the whole of the direct beam is allowed to enter the detector. In making absolute measurements it is usual to employ crystal-reflected radiation so that the intensity determined when the direct beam enters the detector is I_0. There may be a correction due to radiations of wavelengths $\lambda/2$, $\lambda/3$ etc. being reflected from the monochromatizing crystal. This correction can be eliminated by reducing the voltage applied to the X-ray tube below that required to excite the second harmonic of the characteristic radiation (16 kV for Cu $K\alpha$). However, when an X-ray tube is operated under normal conditions and a bent quartz crystal is used as monochromator the increase in the diffuse intensity due to second and higher harmonics does not exceed a few per cent.

The ratio I_d/I_0 is small, often about 10^{-6}, and considerable difficulty is therefore experienced in measuring it. The intensity of the direct beam may be reduced in a number of ways though each method has its own disadvantages. Absorbing screens of thin metal foils may be employed to reduce the intensity by a known factor. If the thickness is t and the linear absorption coefficient is μ then the reduction factor is $\exp(-\mu t)$. There is an experimental uncertainty in the value of μ and since

$$\frac{d(I_d/I_0)}{d\mu} = -te^{-\mu t}$$

$$\frac{d(I_d/I_0)}{I_d/I_0} = -\mu t\frac{d\mu}{\mu}.$$

If the reduction of intensity by the screen is a thousand-fold then $\mu t = 6{\cdot}9$ and hence a 1 per cent. error in the measurement of μ will result in an error of nearly 7 per cent. in I_d/I_0. There is thus a severe limitation on the amount of reduction which can be made by means of absorbing screens. A further difficulty arises with absorbers, owing to second and higher harmonics of the characteristic radiation. As the characteristic radiation is reduced in intensity the number of counts due to $\lambda/2$, $\lambda/3$, etc. radiations increases relative to that due to the λ-radiation. This effect may be small for $\lambda/2$-radiation but is considerable if an appreciable quantity of $\lambda/3$-radiation is present in the incident beam.

The intensity of I_0 may also be reduced by a slotted rotating disk. If the disk has an opening which subtends an angle of $2\pi/100$ at its centre, the intensity is reduced to 1 per cent. in passing through the rotating disk. If the output of X-rays is pulsating it is necessary to be careful to ensure that stroboscopic effects are absent. It is easy to test for the presence of such effects by slightly changing the speed of the disk and noting if any change is registered in the transmitted intensity.

An alternative to the above methods of reducing the intensity of the direct beam to a conveniently measurable quantity is afforded by the use of an intermediate standard. In studying the diffuse reflection around a given relp it is often possible to use the integrated Bragg reflection as such a standard. The measurement is then conducted in two stages. First the ratio between the diffuse intensity associated with a given point in reciprocal space and the integrated intensity of the Bragg reflection is determined. Strictly speaking, it is the integrated intensity of the Bragg reflection, divided by the angular velocity with which the crystal is rotated through the Bragg setting, which is compared with the diffuse intensity. Secondly, the direct beam is compared with the Bragg integrated reflection. The direct beam for this observation is produced by an X-ray tube run at a sufficiently low voltage so that even the second harmonic of the characteristic radiation is not generated. Under these conditions absorbing screens may be used to reduce the incident intensity by a factor of 1000. Proportional counters and scintillating counters can be used so as to discriminate against the second and higher harmonics. When such counters are used it is not necessary to lower the voltage applied to the X-ray tube.

Another intermediate standard can be provided by the intensity of Compton scattering from diamond or other materials containing elements of low atomic number. The whole of the incident X-ray beam is allowed to fall on a face of a diamond crystal set at any convenient angle to the incident beam. The radiation of modified wavelength scattered in directions well away from any Bragg reflections has an intensity which may be calculated from fundamental constants and may be compared directly with the thermal diffuse scattering under investigation. We may define a 'Compton diffuse scattering power', D_c, in an analogous manner to the first order thermal Diffuse Scattering Power, i.e. as the ratio of the intensity scattered by the Compton effect per unit cell of the crystal per unit solid angle to that scattered by a single Thomson electron under the same conditions. For diamond this Compton diffuse intensity, D_c, may be written (James 1948)

$$D_c = 8(6-\sum f_{ec}^2)/B^3,$$

where $\sum f_{ec}^2$ is a function which describes the effect of the individual carbon atoms scattering incoherently. Compton and Allison (1935) give a table (Table II, p. 782) listing the values of $\sum f_{ec}^2$ as a function of the angle of deviation, ϕ, of the X-rays. B is the Breit-Dirac correction factor, the value of which is

$$B = 1+\frac{2h\lambda}{mc}\cdot\frac{\sin^2\theta}{\lambda^2}\,.$$

From this calculated Compton diffuse intensity, D_c, we may determine the absolute intensity of the incident beam, I_0, using the measured intensity of the diffuse beam, I_c, and a formula analogous to that of equation (1.21), namely,

$$\frac{I_c}{I_0} = \frac{\varepsilon^2 D_c \Omega}{2\mu\tau}. \tag{1.22}$$

Such a standardization of the incident intensity can be made with an accuracy of a few per cent.

1:3.2.8. *Monochromatization of the X-ray beam*

For certain types of measurement in connection with the study of diffuse reflections, it is almost essential to use crystal-reflected radiation, or radiation as monochromatic as possible. If white radiation is present in the X-ray beam it is impossible to make measurements along the rel-vector. The intensity of diffuse reflections given by points along rel-vectors of simple indices such as [100] or [110] are often determined by only one or two of the elastic constants. It is therefore necessary to be able to work along the rel-vector as well as along lines on either side of it. The crystals often used for this purpose are quartz, lithium fluoride, and aluminium. There is usually an advantage in bending the crystal either elastically or plastically and curving the surface by grinding to a radius of curvature which is half the radius of bending. It is necessary to limit the divergence of the crystal-reflected beam to that angle, usually not more than 30′, which corresponds to the divergence due to the finite width of the entry slit in front of the detector. For this reason it is not possible to utilize the powerful 002 reflection of pentaerythritol because the divergence of the reflected rays is too great. On the other hand, a plane quartz crystal cut parallel to (10.1) gives a beam which is too well defined and also not as strong as that given by a curved crystal of the same kind. As the curvature of a plane crystal is progressively increased, the intensity of the monochromatic reflected beam also increases. This increase is roughly in proportion to the increase in the divergence of the beam.

An alternative to the monochromatizing crystal which is sometimes used is the system of balanced filters (Chipman and Warren, 1950). Tanaka *et al.* (1959) gave a detailed account of this method. The radiation from a copper anode was passed through a filter of cobalt of thickness corresponding to 10 mg/cm^2 and the intensity of the transmitted $K\alpha$ component was 1·5 per cent. of that incident upon it. A similar filter of nickel transmitted about 60 per cent. Thus if readings are taken first with one filter and then with the other the result is almost entirely due to the monochromatic component. The filters are precisely matched so that they transmit the same proportions of the radiation on either side of the small spectral range which contains the $K\alpha$ line of the copper radiation. The advantage of this method of monochromatization is that the intensity of the monochromatic beam is much higher than it is with crystal-reflected radiation. Unfortunately, there is a small portion of the white radiation which is transmitted more by one filter than by the other, but this usually only involves a small correction to the results. Walker (1956) used both a toroidally bent LiF monochromator and also balanced Ni-Al filters. Extensive accounts of the details of monochromators may be found in standard textbooks, e.g. Guinier (1945*a*), Peiser *et al.* (1955), and Umanskii (1960).

II

THE DETERMINATION OF ELASTIC CONSTANTS OF CRYSTALS BY DIFFUSE X-RAY REFLECTIONS

2:1 Introduction

CLASSICAL methods of determining the elastic constants of crystals fall into two groups. The older methods involve static extension, bending, or twisting of plates, bars, or rods. In the later dynamic methods crystal rods or plates were made to vibrate and the frequency and mode of vibration were correlated with the size, shape, and elastic constants of the crystal. The classical theories of crystal elasticity form a large subject which was almost completely worked out several decades ago. These classical treatments are based on the fundamental assumption that the mechanical equilibrium of a small element of volume is determined only by the forces acting on its surface. All stresses distributed uniformly throughout each volume element are assumed to be without influence on the elastic properties. Within the last decade this assumption has been challenged from both a theoretical and an experimental point of view (Laval, 1951; le Corre, 1953; Viswanathan, 1954; Raman, 1955; Joel and Wooster, 1960). However, it is still too early to be sure that the classical theory must be modified and it seems probable that even if it has to be changed the change of the elastic constants required by the new theory will be small. We shall therefore accept the classical theory as the basis of our discussion.

The classical dynamic method of finding elastic constants may be illustrated by a simple example. When elastic waves are travelling along the tetrad axis of a cubic crystal they are transmitted either as longitudinal or as transverse vibrations. The velocity of travel of the longitudinal wave, V_l, is related to the density, ρ, and the elastic constant, c_{11}, by the equation

$$V_l = \sqrt{\left(\frac{c_{11}}{\rho}\right)}.$$

Thus measurements of the velocity of propagation and of the density give one of the elastic constants. The velocity is related to the frequency, ν, and to the wavelength, Λ, by the equation

$$V_l = \nu\Lambda.$$

Thus the experimental problem resolves itself into the determination of the frequency and wavelength of the waves. In the classical dynamic methods the frequency is imposed by a piezoelectric oscillator or is simply the natural vibrational frequency of a plate or bar. The wavelength is usually determined by the physical dimensions of the vibrating rod or plate and can be measured directly.

A method due to Schaefer and Bergmann (1934, 1935; Bergmann, 1938), has many points of similarity with that to be discussed here. In this method a crystal block is excited piezoelectrically and optical diffraction is made to occur from the travelling elastic waves. The crystal takes the place of the mask E in the optical diffractometer shown in Fig. 1.1. The angle of deviation of the light corresponding to the longitudinal vibration is related to the wavelength of the elastic waves by the usual diffraction formula. When monochromatic light is used the diffraction pattern registered on a photographic film placed at right-angles to the incident light consists of spots distributed along two or three curves. In this experiment the frequency is given by the setting of the piezoelectric oscillator, the wavelength is measured by the displacement of the diffracted image from the centre of the diffraction pattern, and hence the velocity can be found.

In the experiments discussed in this chapter the elastic waves are thermally excited and no mechanical or piezoelectric vibrator is used to produce them. The atoms composing any crystal are always in a state of vibration and owing to the mutual attractions and repulsions of neighbouring atoms their movements can be regarded as due to the superposition of a large number of waves. The wavelengths, frequencies, and wave normal directions of these waves are very numerous though not infinitely variable. When X-rays are allowed to fall upon a crystal each wave train produces its own scattering of the X-rays. There are so many wave trains involved in this scattering that the corresponding diffraction pattern is a cloud, rather than two or three well defined curves, surrounding the normal diffraction spot. However, we can restrict our attention to particular points within this cloud which correspond to wave trains of given frequency and wavelength.

An important difference between the optical and the X-ray diffraction arises from the fact that the optical diffraction pattern occurs only round about the central undeviated beam whereas the diffuse X-rays travel in directions which are close to those of the normal Bragg-Laue reflections and may be inclined at large angles to the incident X-ray beam.

In addition, a perturbed lattice may give rise to diffraction near to the direction of the incident beam, but we shall not be concerned with that in this chapter. The measurement of the wavelength of the elastic waves responsible for a given diffuse beam can thus be measured directly from the settings of the crystal and the detector of the X-rays. The frequency is less easily determined. In the optical diffraction method the frequency is given by the setting of the dials on a piezoelectric oscillator. The elastic waves utilized in the X-ray diffraction are those generated by the thermal vibrations of the atoms. In Chapter I we saw that a longitudinal wave travelling along the tetrad axis, [100], of a cubic crystal gave rise to satellite reflections on either side of each normal reflection. From the geometry of the reciprocal lattice and the reflecting sphere we can, as described in Chapter I, find the displacement **K*** from the relp. If **K*** is directed along [100] then

$$\frac{K^*}{a^*} = \frac{a}{\Lambda},$$

where a and a^* are the unit cell dimensions of the Bravais and reciprocal lattices respectively and Λ is the wavelength of the elastic waves giving rise to the measured diffuse reflection. Thus Λ can be found by measuring K^*.

To determine ν we make use of thermodynamical reasoning, and a measurement of the intensity of the diffuse X-ray reflection. At room temperature the elastic vibrations in most crystals are in equilibrium with one another. In other words, energy is distributed among the various modes of vibration according to the usual thermodynamic laws and each mode carries an energy per unit mass equal to kT, where k is Boltzmann's constant and T is the absolute temperature. This energy is also proportional to the square of the product of the amplitude, ξ, and the frequency. This relationship leads to the equation

$$2\pi^2\xi^2\nu^2 = kT.$$

Thus to find ν we must measure ξ. The intensity of the diffuse reflection is proportional to the square of the product of the amplitude of the elastic wave and the electron density wave corresponding to the nearest relp, as shown in section 1:2.2.2. In equation (1.8) we have the following relation where **K*** and **H*** are both directed along [100]:

$$I(\mathbf{H}^*) = 2\pi^2|F_T|^2(\boldsymbol{\xi}.\mathbf{H}^*)^2.$$

The measurement of $I(\mathbf{H}^*)$ makes it possible to find $\boldsymbol{\xi}$ since the structure amplitude $|F_T|$ and the reciprocal distance of the reflecting planes **H*** are well known. Finally, all the necessary data has been collected and the stages may be summarized by the following equations:

$$\xi^2 = \frac{I(\mathbf{H}^*)}{2\pi^2|F_T|^2H^{*2}} \qquad \Lambda = \frac{a\,.\,a^*}{K^*}$$

$$\nu^2 = \frac{kT}{2\pi^2\xi^2} \qquad V_l = \Lambda\nu$$

$$c_{11} = \rho V_l^2.$$

The outline of the method given above is based on the work of many authors. A short description of the early work leading to the various applications of diffuse X-ray reflections has been given in Chapter I. Important developments followed the papers by Laval (1938, 1939), and Mauguin and Laval (1939). The theoretical treatment involving the scattering of X-rays by thermally excited waves was given by these authors and the essential correctness of the theory was shown by Laval's experimental measurements on sylvine. Further papers developed the theoretical treatment (Laval, 1954*a*, *b*, 1959). Zachariasen (1940), developed independently and almost at the same time, a similar theory, some aspects of which were tested by Siegel and Zachariasen (1940), and Siegel (1941). This treatment was also included in a book by Zachariasen (1945). A different theoretical approach was made by Born and his co-workers: Born (1942*a*, *b*), Sarginson (1942), Pope (1949). The mathematical analysis involved in this work has proved a barrier to its wide use but some authors have used this alternative approach. Lonsdale and Smith (1941) made a careful study by photographic means of various types of diffuse X-ray reflection. Lonsdale (1942*b*, *c*) related diffuse X-ray reflections to the dynamics of crystal lattices. Lifshits (1948) also gave a theory of the scattering of short elastic waves by a crystal lattice. James (1948) dealt with the subject thoroughly in his book, to which many references are made in the present work. A detailed account of many experimental problems and methods arising in connection with this type of study was published by Lonsdale (1948). Since this time many crystals have been studied by the methods discussed in this chapter but the basic ideas have remained unchanged. The diffuse X-ray reflection method is not an easy method of finding the elastic constants of crystals but it is often the only one available when (*a*) the crystals are very small, (*b*) the crystals are soft, (*c*) the measurements must be made at high or low temperatures.

The elastic constants of the following crystals have been determined by diffuse X-ray reflections: KCl (Laval, 1939), Al (Olmer, 1948), KBr, KCl, $NaClO_3$, PbS, $C_6H_{12}N_4$ (Hexamethylene tetramine) (Ramachandran and Wooster, 1949, 1951*b*), ZnS, zincblende (Prince and Wooster, 1951), alpha iron (Curien 1952*a*, *b*), LiF (Hoerni and Wooster, 1952*b*), diamond (Prince and Wooster, 1953), Cu (Jacobsen, 1955), Si (Prasad and Wooster, 1955*a*), Ge (Prasad and Wooster, 1955*b*), β-Sn (Prasad and Wooster, 1955*d*) Al (Walker, 1956; Annaka, 1956), Pb (Prasad and Wooster, 1956*a*), FeS_2, iron pyrites (Prasad and Wooster, 1956*b*), β-AuZn (Schwartz and Muldawer, 1958), V (Sándor and Wooster, 1959).

2:2 Elastic waves in crystals

2:2.1 *Fundamental relations*

The passage of elastic waves through a crystal will be treated here on the classical theory often associated wit theh name of Voigt (1910). Using a more modern notation this theory can be developed as follows. The equation for the displacement, u, of a unit element of volume vibrating under the influence of a wave may be written

$$\rho \frac{\partial^2 u}{\partial t^2} = c \frac{\partial^2 u}{\partial x^2}, \tag{2.1}$$

where ρ is the density and c the elastic constant. This equation expresses on the left-hand side the product of mass and acceleration and on the right the elastic force acting on the volume element.

This expression may be applied to an anisotropic crystal in the following way. The components of the displacement may be denoted $u_m (m = 1, 2, 3)$, where

$$u_m = r_{lm} x_l \tag{2.2}$$

and r_{lm} are the components of strain and x_l are the coordinates of the point suffering the displacement u_m. (Suffixes which are repeated on either side of the equation are to be given the values 1, 2, and 3; those which are not repeated take only one of these values.)

The components of stress t_{pq}† are related to the components of strain r_{lm} by the elastic constants c_{pqlm} according to the generalized form of Hooke's Law, as follows:

$$t_{pq} = c_{pqlm} r_{lm}. \tag{2.3}$$

From equation (2.2) we may write

$$\frac{\partial u_m}{\partial x_l} = r_{lm}$$

(where m and l now have particular values).
Substituting this value of r_{lm} in (2.3) we obtain

$$t_{pq} = c_{pqlm} \frac{\partial u_m}{\partial x_l}. \tag{2.4}$$

The force acting in a given direction on a unit volume element is equal to the rate of change of the stress components in that direction. For example, if $p = q = 1$,

$$\frac{\partial t_{11}}{\partial x_1} = \text{resultant force acting in the direction of the } x_1\text{-axis due to forces normal to the } x_1\text{-axis.}$$

If $p = 1, q = 2$ (Fig. 2.1), then

$$\frac{\partial t_{21}}{\partial x_2} = \text{resultant force acting in the } x_1\text{-direction, due to tangential forces acting on the faces normal to the } x_2\text{-axis.}$$

Thus, in general

$$\frac{\partial t_{qp}}{\partial x_q} = \text{resultant force acting in the } x_p\text{-direction due to stress component } t_{qp}$$

$$= \rho \frac{\partial^2 u_p}{\partial t^2}.$$

† The notation used here is the same as that used in *Textbook on Crystal Physics* by W. A. Wooster (1942) Cambridge University Press (see Appendix III).

The tensors of stress and strain are taken to be symmetrical and we may therefore write, using equation (2.4),

$$\frac{\partial t_{pq}}{\partial x_q} = c_{pqlm}\frac{\partial^2 u_m}{\partial x_q\,\partial x_l} = \rho\frac{\partial^2 u_p}{\partial t^2}. \tag{2.5}$$

If the simple harmonic wave motion has a wave vector **K***† and a frequency ν, we may write

$$u_p = \xi_p e^{2\pi i(\mathbf{K}^*.\mathbf{r}-\nu t)}, \tag{2.6}$$

where **r** is the vector from the origin to the point at the centre of the volume element, and ξ_p is the pth component of the amplitude vector ξ.

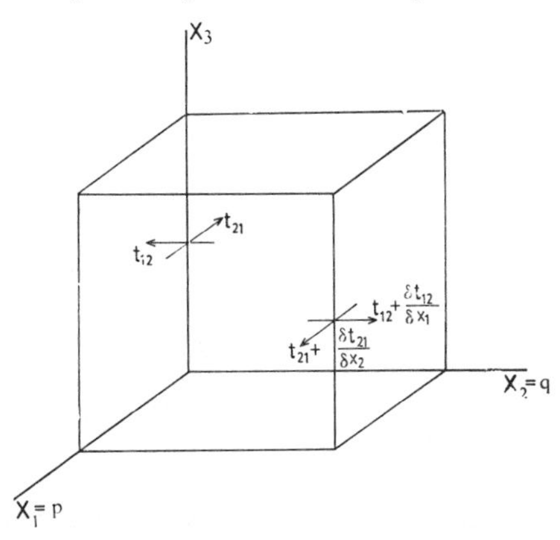

FIG. 2.1 Diagram giving the subscripts of the components of the stress tensor.

In this analysis we seek to find the velocity with which elastic waves travel in directions radiating from a given origin. We can therefore restrict **r** to the same direction as **K***.

From equation (2.6),

$$\frac{\partial u_p}{\partial t} = -2\pi i\nu u_p$$

and

$$\frac{\partial^2 u_p}{\partial t^2} = -4\pi^2\nu^2 u_p. \tag{2.7}$$

Now $\mathbf{K}^* = (f_1\mathbf{n}^*+f_2\mathbf{o}^*+f_3\mathbf{p}^*)|K^*|$ where f_1, f_2, f_3 are the direction cosines of **K*** and **n***, **o***, and **p*** are orthogonal unit reciprocal vectors. Also $\mathbf{r} = x_1\mathbf{n}+x_2\mathbf{o}+x_3\mathbf{p}$, where x_1, x_2, x_3 are the coordinates of the point

† **K*** is normal to the wave front and also to some crystallographic plane of indices *hkl* and therefore may be referred to the reciprocal lattice. This is the reason for the use of the asterisk.

at the end of the vector **r**, and **n**, **o**, and **p** are orthogonal unit vectors in Bravais space corresponding to **n***, **o***, **p***. (It should be noted that this use of a cubic reference system in no way limits the application of the analysis to the cubic system.)

Hence
$$\mathbf{K}^*.\mathbf{r} = (f_1x_1+f_2x_2+f_3x_3)|K^*|$$

and
$$\frac{\partial(\mathbf{K}^*.\mathbf{r})}{x_q} = |K^*|f_q.$$

Differentiating equation (2.6) we obtain

$$\frac{\partial u_m}{\partial x_q} = 2\pi i|K^*|f_q u_m$$

$$\frac{\partial^2 u_m}{\partial x_q\,\partial x_l} = -4\pi^2|K^*|^2 f_q f_l u_m. \tag{2.8}$$

Combining equations (2.5), (2.6), (2.7), and (2.8) we obtain

$$\rho v^2 \xi_p = c_{pqlm} f_q f_l |K^*|^2 \xi_m. \tag{2.9}$$

The quantities $c_{pqlm}f_q f_l$ are all determined by the elastic constants and the direction cosines of the wave normal. We may abbreviate the expressions by writing

$$A_{pm} = c_{pqlm} f_q f_l. \tag{2.10}$$

It is sometimes convenient to have the explicit formulation of A_{pm} and this is given below.†

$$\left.\begin{aligned}
A_{11} &= c_{11}f_1^2+c_{66}f_2^2+c_{55}f_3^2+2c_{56}f_2f_3+2c_{15}f_3f_1+2c_{16}f_1f_2\\
A_{22} &= c_{66}f_1^2+c_{22}f_2^2+c_{44}f_3^2+2c_{24}f_2f_3+2c_{46}f_3f_1+2c_{26}f_1f_2\\
A_{33} &= c_{55}f_1^2+c_{44}f_2^2+c_{33}f_3^2+2c_{34}f_2f_3+2c_{35}f_3f_1+2c_{45}f_1f_2\\
A_{12} &= c_{16}f_1^2+c_{26}f_2^2+c_{45}f_3^2+(c_{25}+c_{46})f_2f_3+\\
&\qquad +(c_{14}+c_{56})f_3f_1+(c_{12}+c_{66})f_1f_2\\
A_{13} &= c_{15}f_1^2+c_{46}f_2^2+c_{35}f_3^2+(c_{36}+c_{45})f_2f_3+\\
&\qquad +(c_{13}+c_{55})f_3f_1+(c_{14}+c_{56})f_1f_2\\
A_{23} &= c_{56}f_1^2+c_{24}f_2^2+c_{34}f_3^2+(c_{23}+c_{44})f_2f_3+\\
&\qquad +(c_{36}+c_{45})f_3f_1+(c_{25}+c_{46})f_1f_2
\end{aligned}\right\} \tag{2.11}$$

† The four suffixes of the c's can be reduced to two because in c_{pqlm}, p and q are interchangeable with one another and so are l and m. In addition,

$$c_{pqlm} = c_{lmpq}.$$

Hence pairs of suffixes are replaceable by single suffixes according to the scheme

11	22	33	23,32	31,13	12,21
1	2	3	4	5	6

2:2.2 *The Christoffel determinant*

If V is the velocity of an elastic wave,

$$V^2 = v^2/|K^*|^2.$$

Equation (2.9) may, therefore, be rewritten

$$\rho V^2 \xi_p = A_{pm}\xi_m. \tag{2.12}$$

On writing this out in full we obtain:

$$\left.\begin{array}{ll} \text{for} \quad p=1, & \xi_1(A_{11}-\rho V^2)+\xi_2 A_{12}+\xi_3 A_{13}=0 \\ \text{for} \quad p=2, & \xi_1 A_{21}+\xi_2(A_{22}-\rho V^2)+\xi_3 A_{23}=0 \\ \text{for} \quad p=3, & \xi_1 A_{31}+\xi_2 A_{32}+\xi_3(A_{33}-\rho V^2)=0 \end{array}\right\}. \tag{2.13}$$

The condition that there shall be a solution of equations (2.13) is that the determinant of the coefficients of ξ_i shall be zero. This determinant is named after Christoffel (1877, 1910), who did so much to advance the study of the elasticity of crystals. This determinant plays a great part in the subsequent development of the subject. From equation (2.10) it is clear that $A_{pm} = A_{mp}$ and hence the Christoffel determinant is symmetrical. The usual form of the equation determining the velocities of the elastic waves is thus

$$\begin{vmatrix} (A_{11}-\rho V^2) & A_{12} & A_{13} \\ A_{12} & (A_{22}-\rho V^2) & A_{23} \\ A_{13} & A_{23} & (A_{33}-\rho V^2) \end{vmatrix} = 0. \tag{2.14}$$

Equation (2.14) is a cubic expression having V^2 as the unknown. The three velocities will be denoted $V_{(1)}$, $V_{(2)}$, and $V_{(3)}$, or, generally, $V_{(i)}$. These wave motions have a common wave normal $\mathbf{K}^*$. Each wave has a polarization vector of unit length giving its vibration direction and these are denoted $\mathbf{e}_{(1)}$, $\mathbf{e}_{(2)}$, and $\mathbf{e}_{(3)}$ respectively. These vectors are parallel to the corresponding amplitude vectors $\boldsymbol{\xi}_{(1)}$, $\boldsymbol{\xi}_{(2)}$, $\boldsymbol{\xi}_{(3)}$ respectively. The components or direction cosines of these polarization vectors are denoted $e_{(i)k}$. The polarization vectors are mutually perpendicular but not necessarily parallel or perpendicular to $\mathbf{K}^*$. One polarization vector, corresponding to a longitudinal wave, is nearly parallel to $\mathbf{K}^*$ and the other two, corresponding to transverse waves, are in general nearly perpendicular to $\mathbf{K}^*$.

As an illustration of the application of this treatment we shall determine the velocities of elastic waves along a [100] axis of a cubic crystal. In this system of symmetry the matrix of elastic constants is as follows:

$$\begin{bmatrix} c_{11} & c_{12} & c_{12} & 0 & 0 & 0 \\ & c_{11} & c_{12} & 0 & 0 & 0 \\ & & c_{11} & 0 & 0 & 0 \\ & & & c_{44} & 0 & 0 \\ & & & & c_{44} & 0 \\ & & & & & c_{44} \end{bmatrix} \tag{2.15}$$

For the wave normal direction [100], $f_1 = 1$, $f_2 = f_3 = 0$, and all $f_q f_l = 0$ except $f_1 f_1 = 1$. Inserting these values into the expression for A_{ik} we obtain

$$A_{11} = c_{1111} = c_{11}, \qquad A_{22} = c_{2112} = c_{44},$$

$$A_{33} = c_{3113} = c_{44}, \qquad A_{ik} = 0, \quad \text{when } i \neq k.$$

Hence the equation (2.14) becomes

$$\begin{vmatrix} c_{11}-\rho V^2 & 0 & 0 \\ 0 & c_{44}-\rho V^2 & 0 \\ 0 & 0 & c_{44}-\rho V^2 \end{vmatrix} = 0. \tag{2.16}$$

The three solutions of this cubic equation are

$$V_{(1)}^2 = \frac{c_{11}}{\rho}, \qquad V_{(2)}^2 = \frac{c_{44}}{\rho} = V_{(3)}^2 .$$

In this example two of the waves capable of travelling in the direction specified have the same velocity. In general there are three distinct solutions to equation (2.14).

The values of the components of the polarization vectors, $e_{(i)k}$, associated with each of the velocities $V_{(i)}$ may now be obtained from equation (2.16). By inserting the value

$$V_{(1)}^2 = \frac{c_{11}}{\rho},$$

we obtain

$$\left.\begin{aligned} e_{(1)1}.0+e_{(1)2}.0+e_{(1)3}.0 = 0 \\ e_{(1)1}.0+e_{(1)2}.c_{44}+e_{(1)3}.0 = 0 \\ e_{(1)1}.0+e_{(1)2}.0+e_{(1)3}c_{44} = 0 \end{aligned}\right\} \tag{2.17}$$

Hence $e_{(1)2} = e_{(1)3} = 0$ and, therefore, $e_{(1)1} \neq 0$. Thus $V_{(1)}$ is the velocity associated with a longitudinal vibration along axis X_1. Similarly, we may obtain for $V_{(2)}$ the values of $e_{(2)k}$, namely, $e_{(2)1} = 0 = e_{(2)3}$ and $e_{(2)2} \neq 0$, and for $V_{(3)}$ we have $e_{(3)k} = 0, 0$, for $k = 1, 2$, and is not zero for $k = 3$.

2:2.3 *Relation between the polarization vectors and the inverse matrix* $(A^{-1})pq$

An important relation exists between certain sums of products of the direction cosines of polarization vectors and the inverse matrix $(A^{-1})pq$. This relation is used later in determining the anisotropy of the thermal diffuse scattering of X-rays.

The nine direction cosines of the vibration directions or polarization vectors are denoted $e_{(i)k}$. If we restrict attention to one of the three waves, say (1), we may write equation (2.12) in terms of direction cosines and obtain

$$e_{(i)p} = A_{pm} \frac{e_{(i)m}}{\rho V_{(i)}^2} \tag{2.18}$$

or, explicitly,

$$e_{(1)1} = A_{11} \frac{e_{(1)1}}{\rho V_{(1)}^2} + A_{12} \frac{e_{(1)2}}{\rho V_{(1)}^2} + A_{13} \frac{e_{(1)3}}{\rho V_{(1)}^2}, \quad \text{etc.} \tag{2.19}$$

Now $e_{(1)1}$, $e_{(2)1}$, and $e_{(3)1}$ are projections of mutually perpendicular unit vectors on to the axis X_1, and hence

$$e_{(1)1}^2 + e_{(2)1}^2 + e_{(3)1}^2 = 1$$

or, in general,

$$\sum_{i=1}^{3} e_{(i)p}^2 = 1. \tag{2.20}$$

Similarly, since $e_{(1)1}$ and $e_{(1)2}$ are the direction cosines of axes X_1, X_2 relative to the polarization vector (1), $e_{(2)1}$ and $e_{(2)2}$ are the direction cosines of the same axes relative to the polarization vector (2), and $e_{(3)1}$ and $e_{(3)2}$ are the direction cosines of the same axes, X_1, X_2, relative to the polarization vector (3), we may write

$$e_{(1)1}e_{(1)2} + e_{(2)1}e_{(2)2} + e_{(3)1}e_{(3)2} = 0$$

or, more generally,

$$e_{(i)p}e_{(i)q} = 0. \tag{2.21}$$

Multiplying equation (2.19) throughout by $e_{(1)1}$, we obtain

$$\left.\begin{aligned}
e_{(1)1}^2 &= A_{11} \frac{e_{(1)1}^2}{\rho V_{(1)}^2} + A_{12} \frac{e_{(1)2}e_{(1)1}}{\rho V_{(1)}^2} + A_{13} \frac{e_{(1)3}e_{(1)1}}{\rho V_{(1)}^2} \\
&\text{and, similarly,} \\
e_{(2)1}^2 &= A_{11} \frac{e_{(2)1}^2}{\rho V_{(2)}^2} + A_{12} \frac{e_{(2)2}e_{(2)1}}{\rho V_{(2)}^2} + A_{13} \frac{e_{(2)3}e_{(2)1}}{\rho V_{(2)}^2} \\
e_{(3)1}^2 &= A_{11} \frac{e_{(3)1}^2}{\rho V_{(3)}^2} + A_{12} \frac{e_{(3)2}e_{(3)1}}{\rho V_{(3)}^2} + A_{13} \frac{e_{(3)3}e_{(3)1}}{\rho V_{(3)}^2}
\end{aligned}\right\} \tag{2.22}$$

Hence, on adding the left-hand sides of the equations (2.22), we have

$$1 = A_{11}\left(\frac{e^2_{(1)1}}{\rho V^2_{(1)}}+\frac{e^2_{(2)1}}{\rho V^2_{(2)}}+\frac{e^2_{(3)1}}{\rho V^2_{(3)}}\right)+$$

$$+A_{12}\left(\frac{e_{(1)1}e_{(1)2}}{\rho V^2_{(1)}}+\frac{e_{(2)1}e_{(2)2}}{\rho V^2_{(2)}}+\frac{e_{(3)1}e_{(3)2}}{\rho V^2_{(3)}}\right)+$$

$$+A_{13}\left(\frac{e_{(1)1}e_{(1)3}}{\rho V^2_{(1)}}+\frac{e_{(2)1}e_{(2)3}}{\rho V^2_{(2)}}+\frac{e_{(3)1}e_{(3)3}}{\rho V^2_{(3)}}\right)$$

or, if i takes the values 1, 2, or 3,

$$1 = A_{11}\frac{e^2_{(i)1}}{\rho V^2_{(i)}}+A_{12}\frac{e_{(i)1}e_{(i)2}}{\rho V^2_{(i)}}+A_{13}\frac{e_{(i)1}e_{(i)3}}{\rho V^2_{(i)}}. \tag{2.23}$$

Similarly, if we multiply $e_{(1)2}$ by $e_{(1)1}$, $e_{(2)2}$ by $e_{(2)1}$, and $e_{(3)2}$ by $e_{(3)1}$, we obtain

$$e_{(1)1}e_{(1)2} = A_{12}\frac{e^2_{(1)1}}{\rho V^2_{(1)}}+A_{22}\frac{e_{(1)1}e_{(1)2}}{\rho V^2_{(1)}}+A_{23}\frac{e_{(1)1}e_{(1)3}}{\rho V^2_{(1)}},$$

$$e_{(2)1}e_{(2)2} = A_{12}\frac{e^2_{(2)1}}{\rho V^2_{(2)}}+A_{22}\frac{e_{(2)1}e_{(2)2}}{\rho V^2_{(2)}}+A_{23}\frac{e_{(2)1}e_{(2)3}}{\rho V^2_{(2)}},$$

$$e_{(3)1}e_{(3)2} = A_{12}\frac{e^2_{(3)1}}{\rho V^2_{(3)}}+A_{22}\frac{e_{(3)1}e_{(3)2}}{\rho V^2_{(3)}}+A_{23}\frac{e_{(3)1}e_{(3)3}}{\rho V^2_{(3)}}.$$

On adding and referring to equation (2.21) we obtain

$$0 = A_{12}\frac{e^2_{(i)1}}{\rho V^2_{(i)}}+A_{22}\frac{e_{(i)1}e_{(i)2}}{\rho V^2_{(i)}}+A_{23}\frac{e_{(i)1}e_{(i)3}}{\rho V^2_{(i)}}. \tag{2.24}$$

Finally, multiplying $e_{(1)3}$ by $e_{(1)1}$, $e_{(2)3}$ by $e_{(2)1}$, and $e_{(3)3}$ by $e_{(3)1}$ and adding, we obtain

$$0 = A_{13}\frac{e^2_{(i)1}}{\rho V^2_{(i)}}+A_{23}\frac{e_{(i)1}e_{(i)2}}{\rho V^2_{(i)}}+A_{33}\frac{e_{(i)1}e_{(i)3}}{\rho V^2_{(i)}}. \tag{2.25}$$

Equations (2.23), (2.24), and (2.25) contain the same three quantities involving products of e's. If these quantities be denoted I, II, III, then we can summarize by writing

$$\left.\begin{aligned} A_{11}\mathrm{I}+A_{12}\mathrm{II}+A_{13}\mathrm{III} &= 1\\ A_{12}\mathrm{I}+A_{22}\mathrm{II}+A_{23}\mathrm{III} &= 0\\ A_{13}\mathrm{I}+A_{23}\mathrm{II}+A_{33}\mathrm{III} &= 0 \end{aligned}\right\} \tag{2.26}$$

From this we may derive the solutions

$$\mathrm{I} = \begin{vmatrix} A_{22} & A_{23} \\ A_{23} & A_{33} \end{vmatrix} \Big/ \Delta = (A^{-1})_{11},$$

$$\mathrm{II} = \begin{vmatrix} A_{23} & A_{12} \\ A_{33} & A_{13} \end{vmatrix} \Big/ \Delta = (A^{-1})_{12},$$

$$\mathrm{III} = \begin{vmatrix} A_{12} & A_{22} \\ A_{13} & A_{23} \end{vmatrix} \Big/ \Delta = (A^{-1})_{13},$$

where Δ is the determinant

$$\begin{vmatrix} A_{11} & A_{12} & A_{13} \\ A_{12} & A_{22} & A_{23} \\ A_{13} & A_{23} & A_{33} \end{vmatrix}.$$

Similarly, we may derive from equations (2.18) the results expressed by the equations

$$\frac{e_{(i)p}e_{(i)q}}{\rho V_{(i)}^2} = (A^{-1})_{pq}, \tag{2.27}$$

where p and q have any one of the values, 1, 2, or 3. Equation (2.27) is of great use in the study of the diffuse reflection of X-rays and determines the anisotropy of the scattering.

2:3 Scattering of X-rays by elastic waves

2:3.1 *The anisotropy of thermal diffuse scattering*

It has been shown in equation (1.8) that the intensity $I(\mathbf{H}^*)$ of the first order thermal diffuse scattering from an elastic wave of amplitude vector $\boldsymbol{\xi}$ and a point in reciprocal space distant $\mathbf{H}^*$ from the origin is proportional to the square of the scalar product of $\boldsymbol{\xi}$ and $\mathbf{H}^*$. Thus

$$I(\mathbf{H}^*) \propto (\boldsymbol{\xi}.\mathbf{H}^*)^2.$$

If we remember that

$$\boldsymbol{\xi} = \xi\mathbf{e}$$

we may write

$$I(\mathbf{H}^*) \propto \xi^2(\mathbf{H}^*.\mathbf{e})^2. \tag{2.28}$$

For most crystals at room temperature the equilibrium between the various modes of elastic vibration requires that the energy contained in each unit cell and associated with a particular mode is equal to kT, where k is Boltzmann's constant and T is the absolute temperature. Now the energy in unit mass of a medium through which a wave of frequency ν and amplitude ξ is travelling is given by $2\pi^2\xi^2\nu^2$. Hence we obtain the relation

$$2\pi^2\xi^2\nu^2 m = kT, \tag{2.29}$$

where m is the mass of the unit cell.† In the expression for the intensity of the first order thermal scattering, (2.28), we may replace ξ^2 by $kT/2\pi^2 v^2 m$. Since there are in general three elastic waves associated with any given wave normal and they each make independent contributions to the total diffuse scattering, we may write for the total diffuse scattering $I(\mathbf{H}^*)$.

$$I(\mathbf{H}^*) \propto \frac{(\mathbf{H}^*.\mathbf{e}_{(1)})^2}{v_{(1)}^2} + \frac{(\mathbf{H}^*.\mathbf{e}_{(2)})^2}{v_{(2)}^2} + \frac{(\mathbf{H}^*.\mathbf{e}_{(3)})^2}{v_{(3)}^2} \qquad (2.30)$$

The point in reciprocal space defined by $\mathbf{H}^*$ also defines $\mathbf{K}^*$, the common wave vector for each of the three waves,

since
$$\mathbf{H}^* = \mathbf{R}^* + \mathbf{K}^*$$

where $\mathbf{R}^*$ is the rel-vector of the nearest relp.
Now

$$v_{(i)} = \mathbf{K}^*.\mathbf{V}_{(i)}, \qquad (2.31)$$

and we may write

$$I(\mathbf{H}^*) \propto \frac{(\mathbf{H}^*.\mathbf{e}_{(i)})^2}{\rho V_{(i)}^2}.$$

We shall put the direction cosines of $\mathbf{H}^*$ equal to those of the rel-vector $\mathbf{R}^*$, namely g_1, g_2, g_3 since $\mathbf{K}^*$ is small in comparison with $\mathbf{R}^*$, and obtain

$$I(\mathbf{H}^*) \propto \frac{(g_m e_{(i)m})^2}{\rho V_{(i)}^2}.$$

On expansion the right-hand expression gives

$$\frac{(g_1 e_{(1)1} + g_2 e_{(1)2} + g_3 e_{(1)3})^2}{\rho V_{(1)}^2} + \frac{(g_1 e_{(2)1} + g_2 e_{(2)2} + g_3 e_{(2)3})^2}{\rho V_{(2)}^2} + $$
$$+ \frac{(g_1 e_{(3)1} + g_2 e_{(3)2} + g_3 e_{(3)3})^2}{\rho V_{(3)}^2}.$$

† This is usually true for points in reciprocal space close to a relp but not accurate enough for points approaching the mid-position between relps. In this case a quantum correction is necessary as described by James (1948) p. 200. The energy of an harmonic oscillator $\bar{E}_{\phi j}$ (using James' symbols) is given by

$$\bar{E}_{\phi j} = (\bar{n}_{\phi j} + \tfrac{1}{2})\, h\omega_{\phi j}$$

where

$$\bar{n}_{\phi j} = 1/(e^{\frac{h\omega_{\phi j}}{kT}} - 1)$$

When kT is much greater than $h\omega_{\phi j}$ this formula gives a value of $\bar{E}_{\phi j}$ very close to kT. When the value of $\bar{E}_{\phi j}$ given by this more accurate formula is inserted in equation 2.29, the frequency $\nu(\omega_{\phi j})$ occurs on both sides of the equation but a solution can be found by successive approximations.

On squaring and collecting terms we obtain

$$I(\mathbf{H}^*) \propto \frac{g_1^2 e_{(i)1}^2}{\rho V_{(i)}^2} + \frac{g_2^2 e_{(i)2}^2}{\rho V_{(i)}^2} + \frac{g_3^2 e_{(i)3}^2}{\rho V_{(i)}^2} + \\ + \frac{2g_1 g_2 e_{(i)1} e_{(i)2}}{\rho V_{(i)}^2} + \frac{2g_2 g_3 e_{(i)2} e_{(i)3}}{\rho V_{(i)}^2} + \frac{2g_3 g_1 e_{(i)3} e_{(i)1}}{\rho V_{(i)}^2},$$

where i takes the values 1, 2, and 3.

Substituting from equation (2.27) we obtain

$$I(\mathbf{H}^*) \propto g_1^2(A^{-1})_{11} + g_2^2(A^{-1})_{22} + g_3^2(A^{-1})_{33} + \\ + 2g_1 g_2(A^{-1})_{12} + 2g_2 g_3(A^{-1})_{23} + 2g_3 g_1(A^{-1})_{31} \tag{2.32}$$

or, more generally,

$$I(\mathbf{H}^*) \propto g_p g_q (A^{-1})_{pq}. \tag{2.33}$$

This equation is the fundamental equation from which the anisotropy of thermal diffuse scattering may be calculated for any crystal.

2:3.2 *K-surfaces*

The variation of the diffuse scattering from points round a given reciprocal lattice point can conveniently be expressed by means of a surface, first used by Jahn (1942*b*), which we shall describe as a K-surface (Prasad and Wooster, 1956*c*). This is defined by equation (2.33) as follows:

$$K(f)_g = g_i g_k (A^{-1})_{ik}, \tag{2.34}$$

where f corresponds to the direction cosines f_1, f_2, f_3 of the wave normal to the elastic wave (rekha), g represents the direction cosines g_1, g_2, g_3 of the line joining the nearest reciprocal lattice point to the origin (rel-vector), and $(A^{-1})_{ik}$ is the inverse matrix derived from the matrix A_{ik}, where

$$A_{ik} = c_{ilmk} f_l f_m. \tag{2.35}$$

Thus the value of $K(f)_g$ is determined simply by the elastic constants and the directions of (*a*) the wave normal corresponding to the source of the diffuse scattering, and (*b*) the rel-vector. In the first place we shall apply this to the cubic system.

The matrix of the elastic constants is as follows:

$$\begin{bmatrix} c_{11} & c_{12} & c_{12} & 0 & 0 & 0 \\ & c_{11} & c_{12} & 0 & 0 & 0 \\ & & c_{11} & 0 & 0 & 0 \\ & & & c_{44} & 0 & 0 \\ & & & & c_{44} & 0 \\ & & & & & c_{44} \end{bmatrix}$$

Hence the values of the terms A_{ik} are:

$$A_{11} = c_{11}f_1^2 + c_{44}f_2^2 + c_{44}f_3^2 = c_{44} + f_1^2(c_{11} - c_{44})$$
$$A_{22} = c_{44}f_1^2 + c_{11}f_2^2 + c_{44}f_3^2 = c_{44} + f_2^2(c_{11} - c_{44})$$
$$A_{33} = c_{44}f_1^2 + c_{44}f_2^2 + f_{11}f_3^2 = c_{44} + f_3^2(c_{11} - c_{44})$$
$$A_{12} = (c_{12} + c_{44})f_1f_2$$
$$A_{13} = (c_{12} + c_{44})f_1f_3$$
$$A_{23} = (c_{12} + c_{44})f_2f_3\,.$$

The corresponding values of $(A^{-1})_{ik}$ are:

$$\begin{aligned}(A^{-1})_{11} &= \begin{vmatrix} A_{22} & A_{23} \\ A_{23} & A_{33} \end{vmatrix} \Big/ \Delta \\ &= (A_{22}A_{33} - A_{23}^2)/\Delta \\ &= \{c_{44}^2 + c_{44}(c_{11} - c_{44})(f_2^2 + f_3^2) + f_2^2f_3^2(c_{11} - c_{44})^2 - \\ &\quad -(c_{12} + c_{44})^2f_2^2f_3^2\}/\Delta \\ &= \{c_{44}^2 + c_{44}(c_{11} - c_{44})(f_2^2 + f_3^2) + \\ &\quad +(c_{11} + c_{12})(c_{11} - c_{12} - 2c_{44})f_2^2f_3^2\}/\Delta,\end{aligned}$$

where Δ is the determinant of the A_{ik} matrix.
Similar expressions may be derived for the remaining five quantities $(A^{-1})_{ik}$. When these are inserted in equation (2.33) we obtain the general expression for a cubic crystal, first given in terms of the indices *hkl* of the relp by Jahn (1942*a*). In the present form this expression is as follows:

$$\begin{aligned}K(f)_g = {} & [g_1^2\{c_{44}^2 + c_{44}(c_{11} - c_{44})(f_2^2 + f_3^2) + (c_{11} + c_{12})(c_{11} - c_{12} - 2c_{44})f_2^2f_3^2\} + \\ & + g_2^2\{c_{44}^2 + c_{44}(c_{11} - c_{44})(f_3^2 + f_1^2) + (c_{11} + c_{12})(c_{11} - c_{12} - 2c_{44})f_3^2f_1^2\} + \\ & + g_3^2\{c_{44}^2 + c_{44}(c_{11} - c_{44})(f_1^2 + f_2^2) + (c_{11} + c_{12})(c_{11} - c_{12} - 2c_{44})f_1^2f_2^2\} - \\ & - 2g_1g_2(c_{12} + c_{44})\{c_{44} + (c_{11} - c_{12} - 2c_{44})f_3^2\}f_1f_2 - \\ & - 2g_2g_3(c_{12} + c_{44})\{c_{44} + (c_{11} - c_{12} - 2c_{44})f_1^2\}f_2f_3 - \\ & - 2g_3g_1(c_{12} + c_{44})\{c_{44} + (c_{11} - c_{12} - 2c_{44})f_2^2\}f_3f_1] \div \\ & \div [c_{11}c_{44}^2 + c_{44}(c_{11} + c_{12})(c_{11} - c_{12} - 2c_{44})(f_1^2f_2^2 + f_2^2f_3^2 + f_3^2f_1^2) + \\ & + (c_{11} + 2c_{12} + c_{44})(c_{11} - c_{12} - 2c_{44})^2f_1^2f_2^2f_3^2].\end{aligned} \tag{2.36}$$

By inserting particular values of the elastic constants and giving f_1, f_2, f_3 all possible values, the whole range of K-values may be covered for any given relp, i.e. for any given set of values of g_1, g_2, g_3. In practice only two or three sets of values of g_1, g_2, g_3 prove to be useful.

For cubic crystals (1,0,0,) and ($1/\sqrt{2}$, $1/\sqrt{2}$, 0) correspond to most requirements, though occasionally $1/\sqrt{3}$, $1/\sqrt{3}$, $1/\sqrt{3}$ may also be useful. The general formula reduces in the particular case when $g_1, g_2, g_3 = (1,0,0,)$ to the following:

$$K(f_1f_2f_3)_{100} = [c_{44}^2+c_{44}(c_{11}-c_{44})(f_2^2+f_3^2)+ \\ +(c_{11}+c_{12})(c_{11}-c_{12}-2c_{44})f_2^2f_3^2]\div \\ \div[c_{11}c_{44}^2+c_{44}(c_{11}+c_{12})(c_{11}-c_{12}-2c_{44})(f_1^2f_2^2+f_2^2f_3^2+f_3^2f_1^2)+ \\ +(c_{11}+2c_{12}+c_{44})(c_{11}-c_{12}-2c_{44})^2f_1^2f_2^2f_3^2]. \tag{2.37}$$

The K-surface can be plotted when particular values for the elastic constants are available. The relative magnitudes of c_{11}, c_{12}, and c_{44} determine the character of this surface though the numerical values depend on the absolute

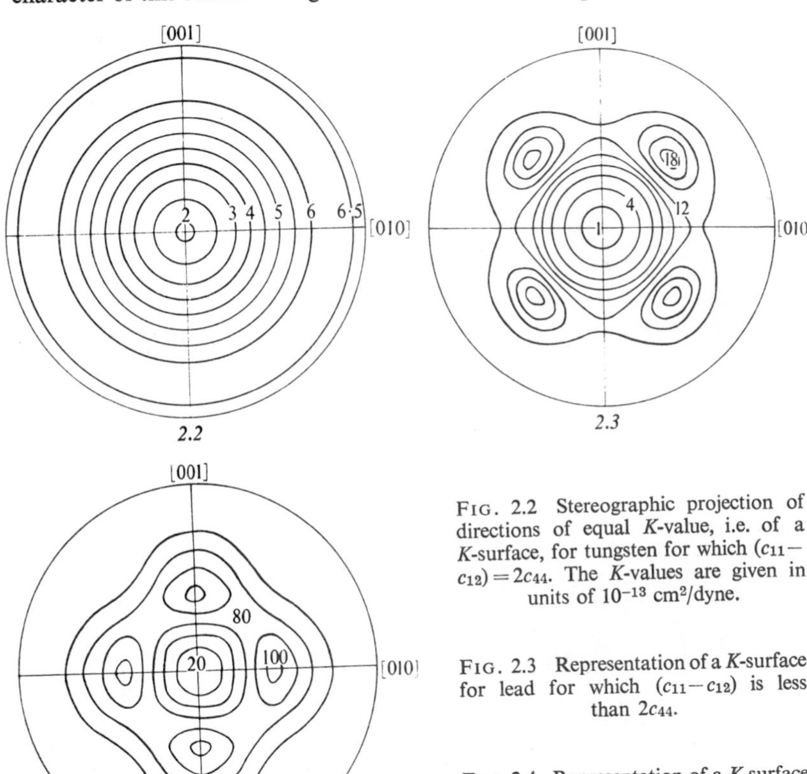

FIG. 2.2 Stereographic projection of directions of equal K-value, i.e. of a K-surface, for tungsten for which $(c_{11}-c_{12})=2c_{44}$. The K-values are given in units of 10^{-13} cm²/dyne.

FIG. 2.3 Representation of a K-surface for lead for which $(c_{11}-c_{12})$ is less than $2c_{44}$.

FIG. 2.4 Representation of a K-surface for which $(c_{11}-c_{12})$ is greater than $2c_{44}$.

values of the constants. In the stereograms which follow, the numbers give $K(f)_g \times 10^{-13}$ and are of dimensions $1/c_{ik}$, i.e. cm^2/dyne.

It will be seen that in the expression for $K(f)_g$ the factor $(c_{11}-c_{12}-2c_{44})$ plays an important role both in the numerator and in the denominator. A single crystal of tungsten has elastic constants such that $(c_{11}-c_{12}-2c_{44})$ is zero. In this case the value of $K(f)_g$ reduces to

$$K(f)_g = \frac{1}{c_{44}} - \left(\frac{1}{c_{44}} - \frac{1}{c_{11}}\right)(g_1 f_1 + g_2 f_2 + g_3 f_3). \tag{2.38}$$

When the wave normal of the elastic wave makes a fixed angle with the rel-vector (which is plotted at the centre of the stereogram), then

$$g_1 f_1 + g_2 f_2 + g_3 f_3 = \text{a constant},$$

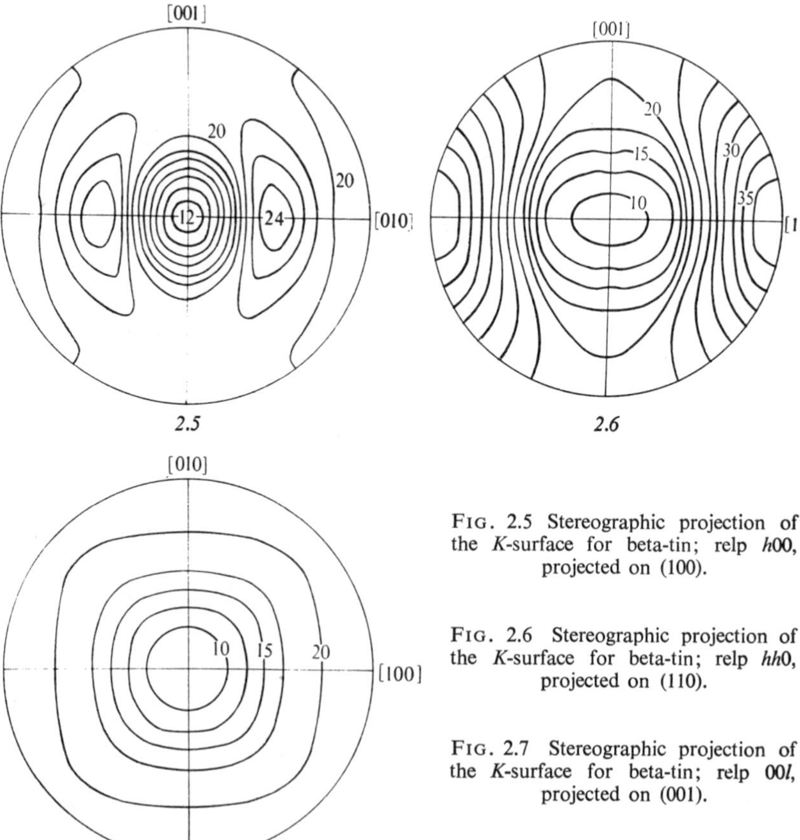

FIG. 2.5 Stereographic projection of the K-surface for beta-tin; relp $h00$, projected on (100).

FIG. 2.6 Stereographic projection of the K-surface for beta-tin; relp $hh0$, projected on (110).

FIG. 2.7 Stereographic projection of the K-surface for beta-tin; relp $00l$, projected on (001).

and hence constant values of $K(f)_g$ correspond to circles about the centre of the stereogram, for every relp. An example of this case is shown in Fig. 2.2, which refers to a single crystal of tungsten.

When $(c_{11}-c_{12})$ is less than $2c_{44}$ we have the case illustrated by lead. Fig. 2.3 is a stereogram of the K-surface about the relp 400. This diagram departs markedly from the cylindrical symmetry shown by tungsten. The other possibility, namely that $(c_{11}-c_{12})$ is greater than $2c_{44}$, is illustrated by Fig. 2.4. This, too, departs from cylindrical symmetry but in a different way from the crystal of lead.

Non-cubic K-surfaces

The expression for $K(f)_g$ in terms of the elastic constants and direction cosines contained in equation (2.34) is applicable to any system of symmetry. The stereograms can be drawn when the elastic constants are known. For beta-tin (tetragonal) Figs. 2.5, 2.6, and 2.7 show the projection of the K-surfaces about the relps $h00$, $hh0$, and $00l$ respectively. As the symmetry decreases more and more elastic constants become involved and the labour of computation increases. However, the K-surface for any relp in a crystal of low symmetry can be calculated if required.

2:3.3 *The absolute magnitude of thermal diffuse scattering*

In sections 2:2.1 to 2:3.2 we have considered the anisotropy of the diffuse scattering, i.e. the relative scattering powers of the disturbed lattice in different directions close to a given Bragg reflection. In this section we shall consider the absolute magnitude of the intensity of diffuse scattering.

The diffuse scattering intensity $I(\mathbf{H}^*)$ in a direction defined by a vector $\mathbf{H}^*$ is given by the equation (1.9), namely,

$$I(\mathbf{H}^*) = 2\pi^2|F_T|^2\sum_{i=1}^{3}(\xi_{(i)}.\mathbf{H}^*)^2,$$

where $|F_T|$ is the modulus of the structure amplitude of the reciprocal lattice point nearest to the point defined by $\mathbf{H}^*$ and $\xi_{(i)}$ is the amplitude vector of the ith wave. If we introduce direction cosines of $\xi_{(i)}$ and $\mathbf{H}^*$, namely $e_{(i)k}$ and g_i respectively, then we may write

$$I(\mathbf{H}^*) = 2\pi^2|F_T|^2\xi^2H^{*2}\sum_{i=1}^{3}(e_{(i)k}g_k)^2. \tag{2.39}$$

If now we substitute from equation (2.29)

$$\xi_{(i)}^2 = \frac{kT}{2\pi^2 m}\cdot\frac{1}{v_{(i)}^2},$$

and from equation (2.31)

$$v_{(i)}^2 = K^{*2}.V_{(i)}^2,$$

we obtain
$$I(\mathbf{H}^*) = |F_T|^2 \frac{kT}{m} \frac{H^{*2}}{K^{*2}} \sum_{i=1}^{3} \frac{(e_{(i)k}g_k)^2}{V_{(i)}^2}$$

or, inserting ρ into numerator and denominator,

$$I(\mathbf{H}^*) = |F_T|^2 \frac{kT}{\tau} \frac{H^{*2}}{K^{*2}} \sum_{i=1}^{3} \frac{(e_{(i)k}g_{(k)})^2}{\rho V_{(i)}^2} \tag{2.40}$$

where τ is the volume of the unit cell.

As we have seen in section 2:3.2. this leads to

$$I(\mathbf{H}^*) = |F_T|^2 \frac{kT}{\tau} \frac{H^{*2}}{K^{*2}} K(f)_g, \tag{2.41}$$

where $K(f)_g$ is the quantity defining the K-surfaces, already discussed in section 2:3.2. As we have seen in section 1:2.2.3, $I(\mathbf{H}^*)$ is known as the first order Diffuse Scattering Power, and is denoted D_1.

Equation (2.41) show that the intensity of diffuse scattering varies as the square of the structure amplitude of the nearest relp, as the square of the length of the rel-vector $\mathbf{H}^*$, and as the inverse square of the wave vector $\mathbf{K}^*$ of the elastic wave producing the reflection. Thus we shall best observe the diffuse scattering near to strong Bragg reflections and more strongly relative to the Bragg reflection, at high θ-values than at low θ-values. The quantity $I(\mathbf{H}^*)$ has throughout been referred to one unit cell of the crystal. The quantity $|F_T|$ is the ratio of the scattering of one unit cell of the crystal to the scattering by one classical electron under the same conditions of angle of scattering and X-ray wavelength. Similarly, $I(\mathbf{H}^*)$ is the ratio of the diffuse scattering from one unit cell to the scattering by one classical electron under the same conditions.

2:3.4 *Second and higher orders of thermal diffuse scattering*

In section 1:2.4 the origin of second and higher order diffuse scattering was discussed. We can now apply the equation (2.40) to determine the magnitude of this effect. In Fig. 1.14 the first order scattering at Q is given by

$$I(\mathbf{OQ}) = |F_T|^2 \frac{kT}{\rho\tau} (OQ)^2 \sum_{i=1}^{3} \frac{(e_{(i)k}g_k)^2}{(AQ)^2 V_{(i)}^2}, \tag{2.42}$$

where $\mathbf{OQ}$ is substituted for $\mathbf{H}^*$ and $\mathbf{AQ}$ for $\mathbf{K}^*$.

The periodicity imposed on the lattice by the waves corresponding to the wave normal $\mathbf{AQ}$ gives a resultant periodicity corresponding to $\mathbf{OQ}$. The periodicity corresponding to the rel-vector $\mathbf{OA}$ has an intensity proportional to $|F_T|^2$, and that corresponding to **OQ** has an intensity proportional to $I(\mathbf{OQ})$. Thus $I(\mathbf{OQ})$ takes the place of $|F_T|^2$ in evaluating second order intensity, $I(\mathbf{OP})^1$, so that we may write

$$I(\mathbf{OP})^1 = I(\mathbf{OQ}) \frac{kT}{\rho\tau} (OQ)^2 \sum_{l=1}^{3} \frac{(e_{(l)m} g_{(m)})}{(QP)^2 V_{(l)}^2}$$

or, inserting the value for $I(\mathbf{OQ})$ given by (2.42) we obtain

$$I(\mathbf{QP}) = |F_T|^2 \frac{k^2 T^2}{2\rho^2 \tau^2} (OQ)^2 (OP)^2 \sum_{i=1}^{3} \sum_{l=1}^{3} \frac{(e_{(i)k} g_k)^2}{(AQ)^2 V_{(i)}^2} \cdot \frac{(e_{(l)m} g_m)^2}{(QP)^2 V_{(l)}^2}. \quad (2.43)$$

The 2 is inserted in the denominator because in carrying out the summation for all values of $\pm$**AQ** and all values of $\pm$**QP** (subject to the requirement that **AQ** + **QP** = **AP**), every term is counted twice. As the whole effect under discussion is small except near to a relp we may replace **OQ** and **OP** by **R***. The detailed evaluation of equation (2.43) is given in Section 3:3 and only the result will be quoted here. It is as follows:

$$D_2 = \frac{\pi^3}{2} \frac{k^2 T^2}{\tau} |F_T|^2 \frac{R^{*4}}{K^*} \sum_{i=1}^{3} \frac{(e_{(i)k} g_k)^4}{\rho^2 V_{(i)}^4} \quad (2.44)$$

Thus the ratio of the second to the first order thermal diffuse scattering is given by

$$I(\mathbf{R}^*)^1 / I(\mathbf{R}^*) = \frac{\pi^3}{2} kT . (R^*)^2 K^* \sum_{i=1}^{3} \frac{(e_{(i)k} g_k)^4}{\rho^2 V_{(i)}^4} \bigg/ \sum_{i=1}^{3} \frac{(e_{(i)k} g_k)^2}{\rho V_{(i)}^2}.$$

In some directions of the vector **OP** it is exactly true that the ratio of the two sigma terms in the above expression is equal to $\mathbf{K}(f)_g$. In other directions the ratio is sufficiently true to enable us to use it in determining the correction to our measurements which must be made on account of the second order thermal diffuse scattering. Finally, we write the ratio as follows

$$I(\mathbf{R}^*)^1 / I(\mathbf{R}^*) = \frac{\pi^3}{2} . kT (R^*)^2 . K^* . K(f)_g . \quad (2.45)$$

The second order contribution to the thermal diffuse scattering seldom amounts to more than ten per cent. of the total effect. A theoretical and experimental study of this matter is due to Paskin (1958). Third order scattering is almost always negligible. Approximate theoretical expressions have been derived for this contribution.

The total diffuse scattering is simply the sum of the first, second, etc., contributions. The second and higher order contributions are estimated and subtracted from the total observed effect. In this way the first order scattering power is derived and from this the elastic constants.

2:4 Examples of the application of the photographic method to the determination of elastic constants

In this section two examples are worked out in detail to show the steps by which by experimental data may be handled. A universal procedure cannot be given because crystals may not exhibit suitable faces, or because X-ray reflections of suitable indices may have too weak an intensity. The ratio of second order scattering to first order scattering is larger in materials of small elastic constants than in materials of large elastic constants. Lead is an example of the former type of crystal and iron pyrites of the latter. Apart from these differences, which are dependent on the crystals examined, there are also differences due to the X-radiation and the type of camera used. It is necessary to avoid producing strong fluorescence in the specimen examined because a general fogging of the film reduces the accuracy of the observations. It is also important to minimize the effect of Compton scattering by a suitable choice of X-ray wavelength. Lead and iron pyrites are very different crystals for our present purposes and represent respectively 'soft' and 'hard' materials.

2:4.1 *Determination of the elastic ratios of iron pyrites*, FeS_2

For a cubic crystal we define two elastic ratios, namely, $\chi_{12} = c_{12}/c_{11}$ and $\chi_{44} = c_{44}/c_{11}$. The values of these two ratios can often be obtained from a single photograph and in combination with a knowledge of the cubic compressibility they permit all three elastic constants to be determined. The following description shows how χ_{12} and χ_{44} were determined for iron pyrites (Prasad and Wooster, 1956*b*). A (110) face was ground on a natural crystal of iron pyrites and lightly etched with concentrated nitric acid. The crystal was mounted on a Unicam S-25 oscillation goniometer with the plane of the (110) face vertical. The radiation used was Mo $K\alpha$ and in front of the film a sheet of zirconium 0·0114 cm thick was placed. The zirconium was necessary to cut out the iron fluorescent radiation. This combination was found more satisfactory than cobalt radiation without any screen before the film. The Bragg angle θ_{440} is 21° 50′ and the angle of incidence i was made equal to 20° 10′ so that $s = +0{\cdot}020$. Starting from the orientation in which the [001] axis was vertical the crystal was rotated in its own plane through 45°, the upward-pointing end of the [001] axis being directed away from the photographic film. The inclination of the rel-vector to the normal to the reflecting sphere, i.e. the angle *NPL* in Fig. 2.8, is 66° 58′. A stereographic plot of the important rekha directions for this orientation of the crystal is given in Fig. 2.9. Using the standard $\bar{\rho}$-, $\bar{\phi}$-, and K^*-charts for $s = +0{\cdot}020$, the positions of the points representing these rekhas were plotted on the microdensitometer contour map and the corresponding K^*-values were read off. This map is drawn by the method described in section 1:3.1.2.

A correction for divergence is necessary for the following reason. The map derived from the photograph would only give the correct distribution of density if the incident beam were infinitely narrow and quite parallel and the crystal were infinitely small. In fact, the Bragg spots are seldom less than 1 × 1 mm in area. Corresponding to the spreading of the spot from point to a finite area the theoretical contours of the intensity distribution are slightly shifted. This displacement can be found as follows. On tracing paper a rectangle, which we shall call the 'integration-area', is drawn corresponding in size with that of a Bragg spot on the actual diffuse reflection photograph. (A Bragg spot of about the same density as the densest part of the diffuse spot must be used in determining this size.) This integration-area is divided into, say, nine equal parts and placed on the contoured experimental intensity distribution. The values of the mean intensity in each of the nine areas are read off and the average of them is found. This average generally differs a little from the value of the intensity at the centre of the integration-area. This process is applied along the particular contours being studied and results in a displacement of the contours from the original positions. Since this integration process is repeating what is done by the actual photograph to the theoretical contour map, we may assume that the contours we seek are on the opposite side of the original experimental contours from those obtained by integration.

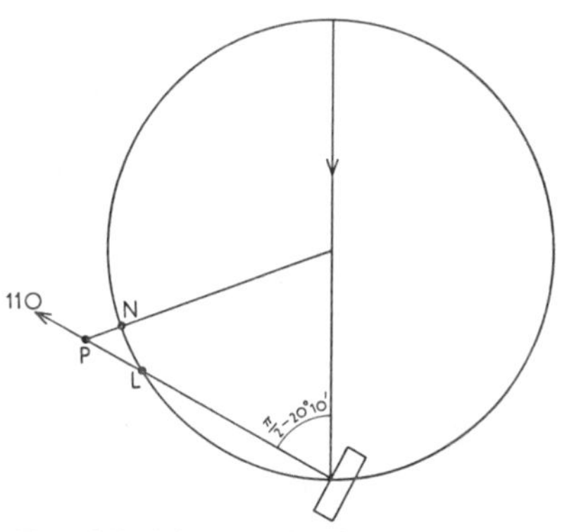

FIG. 2.8 Diagram showing the geometry of the diffuse reflection associated with the 110 relp of iron pyrites.

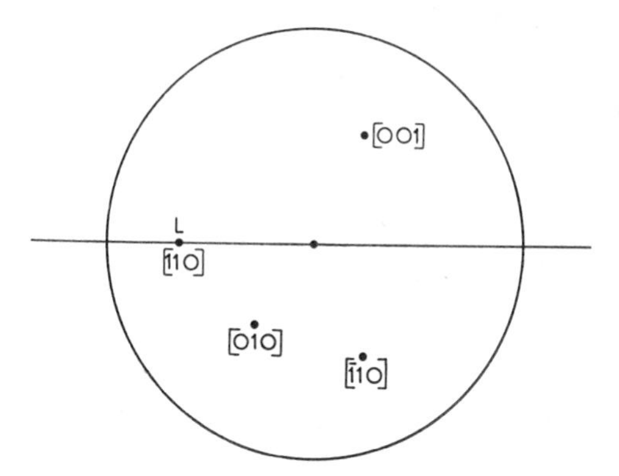

FIG. 2.9 Stereogram showing the important directions of rekhas parallel to [010], [001], and [$\bar{1}$10] in the diffuse reflection for the relp 110 of iron pyrites.

and that the displacements are of equal amount. To sum up the procedure—the experimental contours are shifted a small distance by integration and the true contours are drawn so that the order of the three curves is 'true', 'experimental', 'integrated', and the gap between 'true' and 'experimental' is everywhere equal to that between 'experimental' and 'integrated'.

In general a correction must be applied for the second order diffuse scattering, but in this case the elastic constants were so large that this correction was neglected. From the corrected contour map the intensities at the points marked on the stereogram of Fig. 2.9 were read off. The data are given in Table 2.1

TABLE 2.1

Intensities of diffuse reflections from iron pyrites

Rekha	*Observed intensity I*	$(K_0^*/K^*)^2$	$I(K^*/K_0^*)^2$	$K(f)_g$
[010]	25	0·54	46·5	$\frac{1}{2}(1+1/\chi_{44})$
[001]	31	0·43	72	$1/\chi_{44}$
$[\bar{1}10]$	19·5	0·41	47·5	$2/(1-\chi_{12})$
($K_0^* = NP$ in Fig. 2.8)				

The value of K^* at each of the specified points was read off from the corresponding chart when it was superimposed on the contour map. The column headed $I(K^*/K_0^*)^2$ gives a value proportional to $K(f)_g$, as may be seen from equation (2.41). The column headed $K(f)_g$ gives the theoretical values for the chosen rekhas obtained from equation (2.36).

From the ratio

$$K[010]_{440}/K[001]_{440} = \tfrac{1}{2}(\chi_{44}+1) = 46.5/72,$$

we obtain $\chi_{44} = 0{\cdot}29$.

From the ratio

$$K[001]_{440}/K[\bar{1}10]_{440} = (1-\chi_{12})/2\chi_{44} = 72/47.5,$$

we obtain $\chi_{12} = 0{\cdot}12$.

Other determinations can be made from various combinations of points in the diagram but the illustration given will serve to show how elastic ratios can be easily obtained from an *hh*0 reflection.

2:4.2 *Determination of the elastic constants of lead*

A diffuse reflection photograph, Fig. 2.10*a* (Prasad and Wooster, 1956*a*), was taken using a (100) face prepared by careful cutting and electro-polishing. The radiation used was Cu $K\alpha$, and an S-25 Unicam oscillation goniometer provided with a cylindrical camera of radius 5·73 cm was employed. The Bragg angle θ_{400} is 38° 38′ and the crystal was set at an angle of incidence i, 1° 50′ less than θ, i.e. $\theta - i = 1° \, 50'$. The axis [001] was set parallel to the vertical oscillation axis of the goniometer. At the end of the exposure (20 hours) the crystal was rotated so as to record (during a few seconds) the 420 reflection on the same photograph. In Fig. 2.10*b* this reflection is on the right-hand side. The second photograph, (*b*), is a rotation photograph designed to

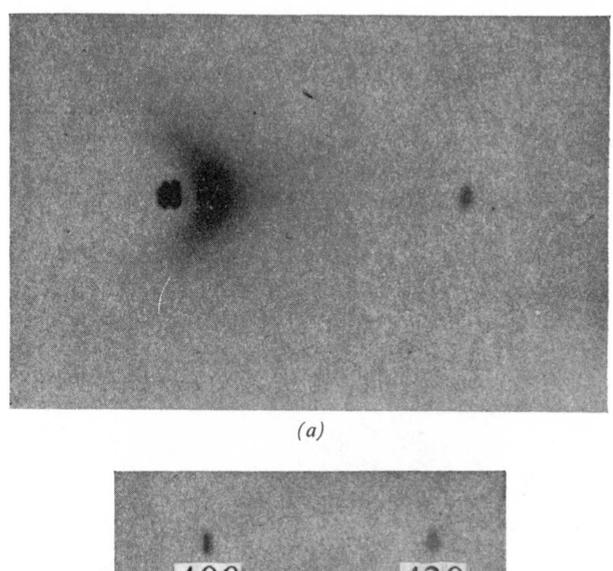

(*a*)

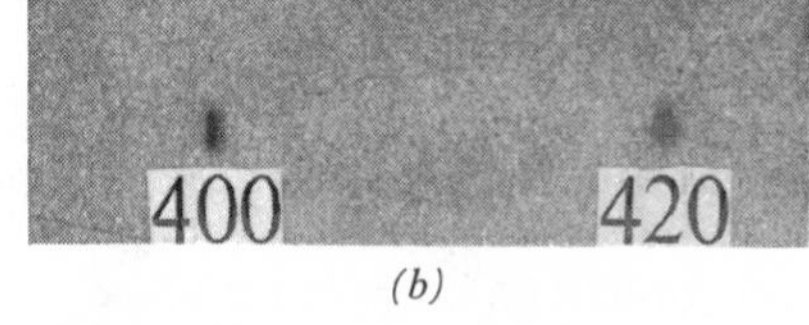

(*b*)

FIG. 2.10*a* Diffuse reflection photograph associated with the 400 relp of a lead single crystal. The axis [001] is vertical.

FIG. 2.10*b* The Bragg reflections from 400 and 420 given by the same crystal when oscillating about the [001] axis.

give the relative position of the Bragg reflections for 420 and 400. By superposing the second photograph on the first so that the 420 reflections were on top of one another, the position of the 400 Bragg reflection relative to the diffuse spot could be determined. The distribution of density over the diffuse spot was measured with an automatic recording microdensitometer. The survey covered an area, at the centre of which was the densest point of the diffuse spot. Traverses parallel to the equatorial line were made, starting from a vertical line 8 mm to the left of the centre of the diffuse spot and going to the same distance on the right-hand side of that point. The traverses across the diffuse spot started 6 mm above the equator and continued to 6 mm below the equator, the separation between succeeding traverses being 0·5 or 1 mm according to the detail of the pattern. From the records of these traverses a contoured map of the distribution of density was obtained. Corrections have to be applied to allow for the finite size of the Laue spot and also for the second order thermal diffuse scattering, which is unusually large in lead. However, these corrections are not necessary in the initial stages of the work and we shall return to them later.

2:4.2.1. *Preliminary determination of* c_{44}/c_{11} *and* c_{12}/c_{11}

The data under discussion are contained in a single diffuse reflection and only the relative values of the diffuse scattering corresponding to different rekhas can be obtained. Our first aim is to obtain a value of c_{44}/c_{11}, i.e. χ_{44}. Either by computation or by a scale drawing (Fig. 2.11), the points L, B, C at

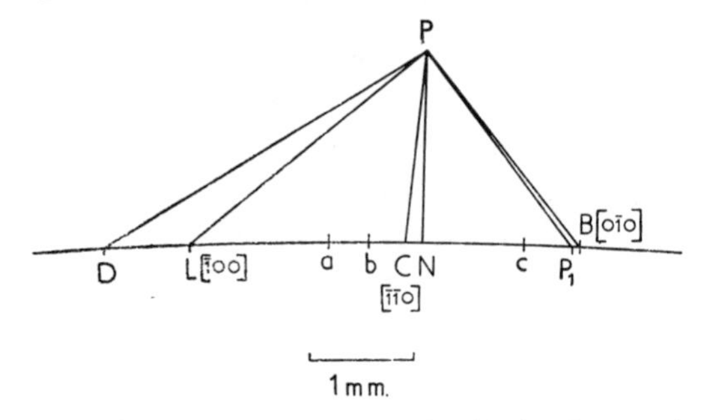

FIG. 2.11 Diagram by means of which the photograph of Fig. 2.10*a* may be interpreted. The letters L, N, P_1, and P have the same significance as in Fig. 1.25.

which the rekhas [$\bar{1}$00], [0$\bar{1}$0], [$\bar{1}\bar{1}$0] intersect the photograph are determined. The scale of the microdensitometer record is such that 1 mm on the original film corresponds to 27 mm on the record. The record therefore corresponds to a photograph taken in a camera of radius $27 \times 5{\cdot}73 = 154{\cdot}7$ cm. We shall assume that the reflecting sphere is also of this radius.

With the values of $\theta = 38° \, 38'$ and $(\theta - i) = 1° \, 50'$, we have from equation (1.16) $\varepsilon = 0{\cdot}025$ and from equation (1.13)

$$s = 2 \sin \theta . (\theta - i) . \cos\left(\theta + \frac{\theta - i}{2} - \frac{\varepsilon}{2}\right) = 0.0311.$$

The distance of the relp P (Fig. 2.11) from the reflecting circle passing through L and B is, therefore, $154{\cdot}7 \times 0{\cdot}0311 = 4{\cdot}81$ cm. In Fig. 2.11 the line LB is an arc of a circle of radius 154·7 cm. This can be drawn, either with a beam compass, or by the following construction. A horizontal line is drawn through N perpendicular to PN. This is a tangent to the reflecting circle at N. At a distance x along the tangent from N the circumference of the circle is a distance below the tangent equal to $x^2/2 \times 154{\cdot}7$ cm within the necessary accuracy. Several points along the circumference can thus be found and a circle drawn through them.

The line PL (see also Figs. 1.25 and 1.26) makes with the line PN an angle $(\pi/2-\theta)+\varepsilon-(\theta-i) = 51°\ 22'+1°26'-1°\ 50' = 50°\ 58'$. PP_1 is an arc of a circle of radius $2 \sin \theta . 154{\cdot}7 = 193{\cdot}2$ cm, and cuts the reflecting circle in the point P_1. P_1 is the position on the diagram which the Bragg spot would occupy if it were present. The line PL is the axis $[\bar{1}00]$ and if we draw lines PB, PC at angles of 90° and 45° respectively with PL, the points B and C correspond to the intersections of the rekhas $[0\bar{1}0]$ and $[\bar{1}\bar{1}0]$ respectively with the reflecting circle.

From the microdensitometer measurements of the distance between the diffuse spot and the Bragg spot 420 (Fig. 2.10*a*) and the distance between the spots 420 and 400 (Fig. 2.10*b*), the position of the point P_1 relative to the diagram of the density distribution can be fixed. Along the equatorial line of the photograph there are three rekhas which are simply related to the elastic constants. These are the $[\bar{1}00]$, $[0\bar{1}0]$, and $[\bar{1}\bar{1}0]$ rekhas for which the expression (2.37) reduces to

$$K[\bar{1}00]_{400} = 1/c_{11}$$

$$K[0\bar{1}0]_{400} = 1/c_{44}$$

$$K[\bar{1}\bar{1}0]_{400} = 1/(c_{11}-c_{12})+1/(c_{11}+c_{12}+2c_{44}).$$

Thus the ratio

$$\frac{K[\bar{1}00]_{400}}{K[0\bar{1}0]_{400}} = \frac{c_{44}}{c_{11}} = \chi_{44}$$

provides a direct determination of the elastic ratio χ_{44}. When X-radiation is used which is filtered, but not crystal-reflected, the Laue spot occurs just where the rekha $[\bar{1}00]$ emerges on to the film. However, the nearest usable point to the Laue spot on the left-hand side gives a K-value which is only a few per cent. different from the value for the [100] rekha itself. From this point an approximate value of χ_{44} can be obtained. The diffuse intensity at the point D (Fig. 2.11) is not affected by the Laue spot and corresponds to 1·7 scale divisions on the microdensitometer record. The intensity of diffuse X-rays is inversely proportional to K^{*2}, i.e. to PD^2. The intensity at the point B corresponds to 15·5 scale divisions and the K^*-value is PB.

Hence

$$\frac{K[\bar{1}00]_{400}}{K[0\bar{1}0]_{400}} = \frac{1{\cdot}7}{15{\cdot}5} \times \frac{PD^2}{PB^2} = \frac{1{\cdot}7}{15{\cdot}5} \times \frac{9{\cdot}70^2}{6{\cdot}32^2} = 0{\cdot}26 = \chi_{44}.$$

Because of the small value of 1·7 scale divisions the accuracy of this measurement is low and this is inevitable with reflections of filtered radiation from (100) faces.

To determine the elastic ratio c_{12}/c_{11}, i.e. χ_{12}, we can proceed in a similar manner, using the intensities at the points corresponding to L and C in Fig. 2.11. The diffuse intensity at the point corresponding to C is 58 and the length $PC = 4{\cdot}87$. Hence, from the expressions for the K-values we have

$$\frac{K[\bar{1}00]_{400}}{K[\bar{1}\bar{1}0]_{400}} = \frac{1{\cdot}7}{58} \times \frac{9{\cdot}70^2}{4{\cdot}87^2} = 0{\cdot}116$$

$$= \frac{1/c_{11}}{1/(c_{11}-c_{12})+1/(c_{11}+c_{12}+2c_{44})}$$

$$= \frac{1}{2}\frac{(1-\chi_{12})(1+\chi_{12}+2\chi_{44})}{(1+\chi_{44})}.$$

If we insert the value $\chi_{44} = 0{\cdot}26$ we obtain $\chi_{12} = 0{\cdot}88$. These are the preliminary values which are used in the refining process described in the following sections.

2:4.2.2 *Correction for second order scattering*

There are several ways in which the correction for second order thermal diffuse scattering may be obtained and the method described here has been found in practice to be the most easy to apply. A theoretical K-surface based on the preliminary values of the elastic ratios χ_{12} and χ_{44} is first obtained. It is laborious to calculate the course of all the contours on such a K-surface but with a preliminary survey of the problem the relevant parts can be found and attention can be concentrated on them. If a standard set of K-surfaces, covering the range of elastic ratios under discussion, is available the appropriate K-surface may be selected. This K-surface is now reorientated to correspond to the actual orientation of the [100], [010], and [001] axes with respect to the surface of the reflecting sphere. Using a Wulff net the K-surface is rotated until the $[\bar{1}00]$ axis makes an angle LPN (50° 58′), Fig. 2.11, with the direction, PN, perpendicular to the paper. The contours of the K-surface are now transferred by means of the $\bar{\rho}$-, $\bar{\phi}$-values read off from the stereogram to a drawing on the same scale as the microdensitometer record. Thus, with the centre of the projection sphere at P, Fig. 2.11, all lines through the contours of the K-surface are projected on to the curved surface corresponding to reflecting sphere LCB. (This projection may be done using a chart constructed for $s = 0{\cdot}0311$ or by using the formulae (1.10), (1.11), and (1.12) (section 1:3.1.4) which relate the Cartesian coordinates of a point on the film to the corresponding $\bar{\rho}$-, $\bar{\phi}$-values.) On this projection the circles are drawn, giving the value of K^* at any point. The intensity of the diffuse scattering at any point is proportional to the product of the K-value and K^{*2} at that point. Taking this product at the centre point (Fig. 2.11) as unity, two contours corresponding to the value of the product equal to 0·5 and 0·3 respectively are traced on the diagram (which is similar to Fig. 2.12). This diagram is the ideal microdensitometer contour diagram which would be obtained with an infinitely small Laue spot and zero intensity of second order thermal diffuse scattering. The second

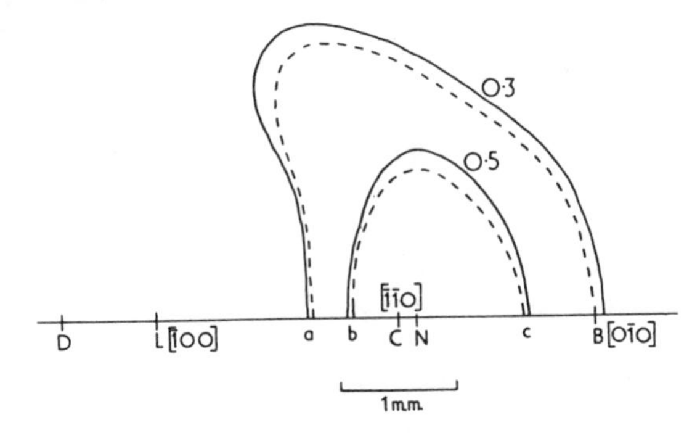

FIG. 2.12 Contoured distribution map of the intensity in the diffuse reflection shown in Fig. 2.10*a*. The continuous lines give the experimentally determined contours and the dotted lines the contours after correction for the second order diffuse scattering and for the finite size of the Bragg spot. (After Prasad and Wooster, 1956*a*).

order thermal diffuse scattering has an intensity $I(R^*)^1$ which can be calculated as a fraction of the first order diffuse intensity by the equation (2.45), namely,

$$I(\mathbf{R}^*)^1/I(\mathbf{R}^*) = \tfrac{1}{2}\pi^3 kT(R^*)^2 K^* K(f)_g. \tag{2.46}$$

For any given K-value this ratio can be evaluated. When applied to all the points along and near to a given contour, such as, for example, that denoted 0·5, the new course of the theoretical 0·5 contour can be traced.

The correction for the finite size of the normal X-ray reflection, and the resulting movement of the contour in such a diagram as Fig. 2.11, are determined as described in section 2:4.1. The displacements of the contours for the two corrections, due respectively to second order scattering and to divergence, are added together. Up to the present we have worked from a theoretical K-surface based on an approximate knowledge of the values of χ_{12} and χ_{44} and have determined by how much the two corrections shift the 0·5 and 0·3 contours. We now apply the same corrections in the opposite sense to the experimentally determined contours in order to obtain the theoretical contours. The continuous line contours of Fig. 2.12 are those determined directly from the photograph and the dotted line contours are those obtained after applying the corrections.

2:4.2.3 *Refinement of the values of c_{44}/c_{11} and of c_{12}/c_{11}*

In Fig. 2.12 the continuous lines represent the contours derived from the microdensitometer records and the dotted lines the corrected contours obtained by the method described above. To illustrate the use of such a diagram in deriving successively more accurate values of the elastic ratios χ_{12} and χ_{44} we shall study the points on the central horizontal line denoted *a*, *b*, and *c* respectively. The point *a* is near to the Laue spot and the intensity at *a* is mainly determined by the value of c_{11}. The intensity at *B* is mainly affected by the value of c_{44} and, lastly, the intensity at point *C* is mainly determined by c_{12}. The following procedure can therefore be adopted.

(1) Combine information concerning points *a* and *B* to improve the value of χ_{44}.
(2) Using this value of χ_{44}, combine information concerning points *b* and *B* to improve the value of χ_{12}.
(3) Combine information concerning points *c* and *B* to obtain a further improved value of χ_{44}.

The data concerning the four points *a*, *b*, *c*, and *B*, $[0\bar{1}0]$, is contained in Table 2.2.

TABLE 2.2

Experimental data obtained from the diffuse reflection photograph of a lead single crystal

Point	$\bar{\rho}$	$\bar{\phi}$	f_1^2	f_2^2	I	$(s/K^*)^2$	$I/(s/K^*)^2$
a	90°	117°	0·8349	0·1651	0·3	0·79	0·38
b	90°	107°	0·6857	0·3144	0·5	0·90	0·56
c	90°	62°	0·0357	0·9644	0·5	0·76	0·66
B, $[0\bar{1}0]$	90°	51°	0·0000	1·000	0·34	0·59	0·58

(1) Point *a*

For the point *a* on the equatorial plane

$$c_{11}K[f_1f_20]_{400} = \frac{\chi_{44}f_1^2+f_2^2}{\chi_{44}+(1+\chi_{12})(1-\chi_{12}-2\chi_{44})f_1^2f_2^2} \tag{2.47}$$

and, using our approximate value $\chi_{12} = 0{\cdot}88$, we obtain for point *a*

$$c_{11}K[f_1f_20]_{400} = \frac{0{\cdot}835\chi_{44}+0{\cdot}165}{\chi_{44}+0{\cdot}259(0{\cdot}12-2\chi_{44})}$$

For the point *B*,

$$c_{11}K[0\bar{1}0]_{400} = \frac{1}{\chi_{44}}.$$

Hence from Table 2.2,

$$\frac{c_{11}K[f_1f_20]_{400}}{c_{11}K[0\bar{1}0]_{400}} = \frac{0{\cdot}38}{0{\cdot}58} = 0{\cdot}655 = \frac{(0{\cdot}835\chi_{44}+0{\cdot}165)\chi_{44}}{\chi_{44}+0{\cdot}259(0{\cdot}12-2\chi_{44})}.$$

The solution of this expression gives $\chi_{44} = 0{\cdot}27$.

(2) Point *b*

We now use this value of χ_{44} in the combination of data for points *b* and *B* to find a better value for χ_{12}. Using the same formula (2.47) and inserting the values of f_1^2, f_2^2 for point *b*, we obtain

$$c_{11}K[f_1f_20]_{400} = \frac{0{\cdot}185+0{\cdot}314}{0{\cdot}25+(1+\chi_{12})(1-\chi_{12}-0{\cdot}50)0{\cdot}215},$$

$$c_{11}K[0\bar{1}0]_{400} = \frac{1}{\chi_{44}} = 3{\cdot}71,$$

and from Table 2.2 the ratio of these two *K*-values is 0·966. Solving the above equation we obtain $\chi_{12} = 0{\cdot}80$.

(3) Point *c*

Finally, we use this corrected value of χ_{12} with the data on point *c* to obtain a further corrected value of χ_{44}, namely 0·26.

Thus, from this photograph we arrive at the values for the elastic ratios χ_{12} and χ_{44} of 0·80 and 0·26 respectively. The values determined by the diffractometer (Section 2:5.1) are rather more accurate, namely 0·78 and 0·28 respectively.

2:4.2.4 *Absolute values of the elastic constants*

From a single diffuse reflection photograph it is possible to obtain the elastic ratios χ_{12} and χ_{44}. To obtain the absolute values of the elastic constants it is necesary to know either the absolute intensity of the diffuse reflecting power or else some physical property, such as the linear or cubical compressibility, which depends on the absolute values of one or more of the elastic constants. The relation between the cubical compressibility β and the elastic moduli and constants is as follows (Wooster, 1938, p. 255):

$$\beta = 3(s_{11}+2s_{12}) = \frac{3}{c_{11}+2c_{12}}$$

or

$$\frac{1}{\beta} = \tfrac{1}{3}c_{11}(1+2\chi_{12}) \qquad (2.48)$$

The compressibility at zero pressure for lead is $23{\cdot}3\times10^{-13}\text{cm}^2/\text{dyne}$ (Bridgman 1938, 1945) and, using the value 0·78 for χ_{12}, we obtain from equation (2.48)

$$c_{11} = 5{\cdot}0(3)\times10^{11}\ \text{dyne/cm}^2.$$

Finally, from the values $c_{12}/c_{11} = 0{\cdot}78$ and $c_{44}/c_{11} = 0{\cdot}28$ we obtain the absolute values of the other two constants, namely,

$$c_{12} = 3{\cdot}9(3), \qquad c_{44} = 1{\cdot}4(0) \times 10^{11} \text{ dyne/cm}^2.$$

2:5 Examples of the application of the diffractometer method to the determination of elastic constants

When a large enough crystal and a sufficiently strong crystal-reflected monochromatic X-ray beam are available, more accurate results can be obtained by a diffractometer method than by a photographic one. It is advisable to take diffuse reflection photographs in any case so as to be sure that genuine thermal effects rather than structural or polyphase phenomena are being studied. The divergence corrections are generally more serious with the diffractometer than with the photograph because, in seeking to increase the rate of counting of the diffusely scattered quanta, the solid angle subtended by the window of the counter at the specimen is made as large as possible. The divergence corrections have been discussed in section 1:3.2.6. The following examples, namely lead and diamond, are chosen because they represent extreme cases of crystals with small and large elastic constants respectively.

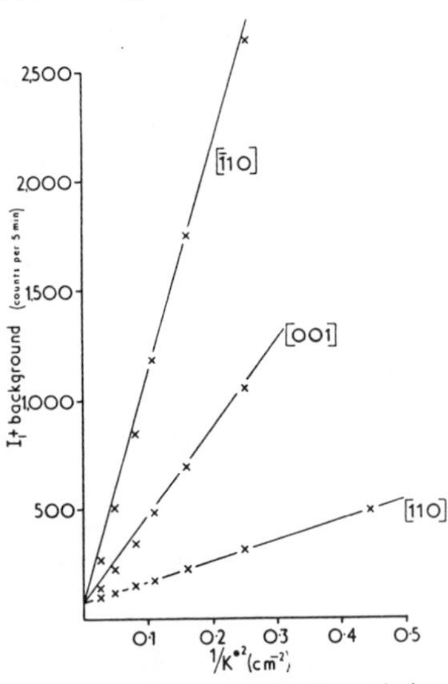

FIG. 2.13 Curves showing the variation of diffuse intensity plotted as a function of $1/K^{*2}$ for lead, using the relp 220 and various rekhas. (After Prasad and Wooster, 1956*a*).

2:5.1 *The elastic constants of lead*

Single crystals grown from the melt were cut parallel to a (110) plane (Prasad and Wooster, 1956*a*). The crystals were very soft and easily distorted by any kind of cutting or mechanical handling. For this reason bending and twisting methods are difficult to use in measuring the elastic properties of lead. Electro-polishing removed the disturbed surface layer and the crystals gave sharp Laue spots. Using Cu $K\alpha$ radiation diffuse reflection in the neighbourhood of the 220 relp was studied. The rekha directions used were [110], [10$\bar{1}$], and [001], and the *i*-, ϕ-charts were used to obtain the appropriate settings of the crystal and detector. Table 2.3 gives the distance from the relp along the rekha at which the diffuse measurement was made. (The radius of the reflecting sphere is taken as 50 cm.) The column headed I_2 gives the

TABLE 2.3

Data on diffuse scattering from lead λ = 1·54 Å

Rekha	K^* (*cm*)	$1/K^{*2}$ (cm^{-2})	*Total Intensity*	I_2	I_1+ *background*	*Mean slope*
$[110]_{220}$	1·5	0·444	495	2·9	492·1	
	2·0	0·250	318·7	2·2	316·5	
	2·5	0·160	233·4	1·8	231·6	
	3·0	0·111	174·6	1·4	173·2	
	3·5	0·0816	153·8	1·3	152·5	
	4·5	0·0498	116·9	0·9	116·0	
	6·0	0·0278	100·4	0·8	99·6	985
$[\bar{1}10]_{220}$	2·0	0·250	2927	289	2638	
	2·5	0·160	1961	234·4	1726·6	
	3·0	0·111	1375·8	191·0	1184·8	
	3·5	0·0816	1007·3	157·6	849·7	
	4·5	0·0498	626·8	117·4	509·4	
	6·0	0·0278	348	77·0	271·0	10,540
$[001]_{220}$	2·0	0·250	1075	19·6	1055·4	
	2·5	0·160	717·8	16·0	701·8	
	3·0	0·111	499·1	12·9	486·2	
	4·5	0·0498	237·8	8·2	229·6	
	6·0	0·0278	154·6	6·3	146·3	4,110

contribution due to the second order diffuse scattering, calculated as described in section 2:3.4 and by equation (2.45). The column 'I_1+background' is plotted against $1/K^{*2}$ for each rekha as shown in Fig. 2.13. From the slopes of the straight lines drawn through the points and a common point on the axis of ordinates the K-values are seen to be in the following ratios:

$$K[110]_{220} : K[\bar{1}10]_{220} : K[001]_{220} = 985 : 10{,}540 : 4{,}110.$$

From the formula (2.36) for the general value of $K(f)_g$ in the cubic system we obtain

$$K[110]_{220} = 2/(c_{11}+c_{12}+2c_{44})$$

$$K[\bar{1}10]_{220} = 2/(c_{11}-c_{12})$$

$$K[001]_{220} = 1/c_{44}.$$

Using these relations and the experimentally determined ratios, we obtain

$$\chi_{12} = 0{\cdot}78(2), \qquad \chi_{44} = 0{\cdot}27(9).$$

The compressibility, β, at zero pressure given by Bridgman (1945), $23{\cdot}3 \times 10^{-13}$ cm^2/dyne.

From the relation

$$\frac{1}{\beta} = \tfrac{1}{3}(c_{11}+2c_{12}) = \tfrac{1}{3}c_{11}(1+2\chi_{12}),$$

we obtain

$$c_{11} = 5{\cdot}0(3), \qquad c_{12} = 3{\cdot}9(3), \qquad c_{44} = 1{\cdot}4(0) \times 10^{11} \text{ dyne/cm}^2.$$

Owing to the large anisotropy shown by lead, for which $K[\bar{1}10]_{220}$ is more than ten times greater than $K[110]_{220}$, the accuracy of this determination is probably less than the ± 5 per cent. which is usual in such measurements.

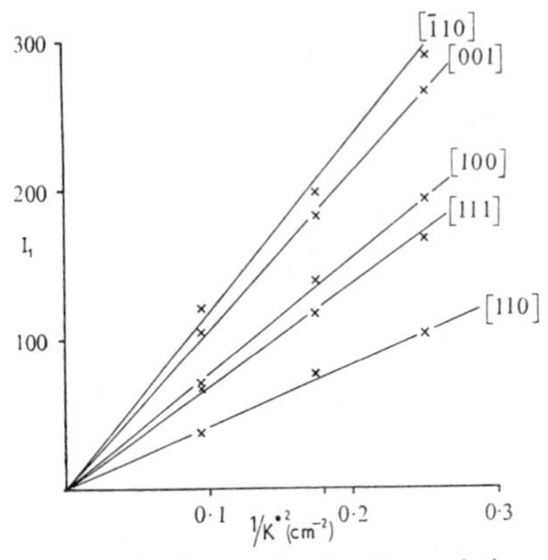

FIG. 2.14 Curves showing the variation of diffuse intensity plotted as a function of $1/K^{*2}$ for diamond, using the relp 220 and various rekhas. (After Prince and Wooster, 1953).

2:5.2 *The elastic constants of diamond*

In contrast with single crystals of lead, diamond is extremely hard and has such large values of the elastic constants that at room temperature there is little thermal diffuse scattering. A crystal with a maximum dimension of 2 cm was cut parallel to the (110) face (Prince and Wooster, 1953) and mounted in a small furnace. This furnace was carried on a goniometer head having two arcs and one slide and could raise the temperature of the diamond to about 320°C. The diffuse intensities along a number of rekhas were determined by the use of Cu $K\alpha$ crystal-reflected radiation. The results are given in Table 2.4. No correction for I_2 is necessary because the elastic constants have such a high value. The results are plotted in Fig. 2.14 and from these straight lines the 'mean slope' given in the last column of Table 2·4 has been derived.

The values of the expressions

$$\log_{10} \frac{K[ABC]_{220}}{K[110]_{220}}$$

for the rekhas [100], [1$\bar{1}$0], [111], and [001] were found to be respectively 0·27, 0·45, 0·22, and 0·41. Using the relations derived from equation (2.36), namely,

$$K[110]_{220} = 2/(c_{11}+c_{12}+2c_{44}),$$

$$K[100]_{220} = \frac{1}{2}\left(\frac{1}{c_{11}}+\frac{1}{c_{44}}\right),$$

$$K[1\bar{1}0]_{220} = 2/(c_{11}-c_{12}),$$

$$K[111]_{220} = 2/(c_{11}+2c_{12}+4c_{44})+1/(c_{11}-c_{12}+c_{44}),$$

$$K[001]_{220} = 1/c_{44},$$

graphs were plotted of the logarithms of the K-ratios for values of χ_{12}, 0·2–0·4, and of χ_{44}, 0·3–0·5. The four lines so obtained are shown in Fig. 2.15, from which it will be seen that the best mean values are $\chi_{12} = 0{\cdot}3$, $\chi_{44} = 0{\cdot}4$.

An absolute measurement of $K[001])_{220}$ was made by the method employing the Compton scattering from diamond, described in section 1:3.2.7. This

TABLE 2.4

Data on diffuse scattering from diamond $\lambda = 1{\cdot}54$ Å

Rekha	K^* (*cm*)	$1/K^{*2}$ (cm^{-2})	*Number of counts/min.* I_1	*Mean slope*
$[110]_{220}$	2·0	0·25	1	
	2·39	0·175		
	3·25	0·095		104
$[100]_{220}$	2·0	0·25	194	
	2·39	0·175	140	
	3·25	0·095	72	193
$[\bar{1}10]_{220}$	2·0	0·25	290	
	2·39	0·175	199	
	3·25	0·095	122	296
$[111]_{220}$	2·0	0·25	168	
	2·39	0·175	119	
	3·25	0·095	68	171
$[001]_{220}$	2·0	0·25	266	
	2·39	0·175	183	
	3·25	0·095	106	267

Background 90 counts/min.

gave a value of $c_{44} = 44 \times 10^{11}$ dyne/cm^2. From this value of c_{44} and the elastic ratios χ_{12} and χ_{44}, the elastic constants were determined as

$$c_{11} = 110 \times 10^{11}, \qquad c_{12} = 33 \times 10^{11}, \qquad c_{44} = 44 \times 10^{11} \text{ dyne/cm}^2.$$

The accuracy of these results is about ± 10 per cent.

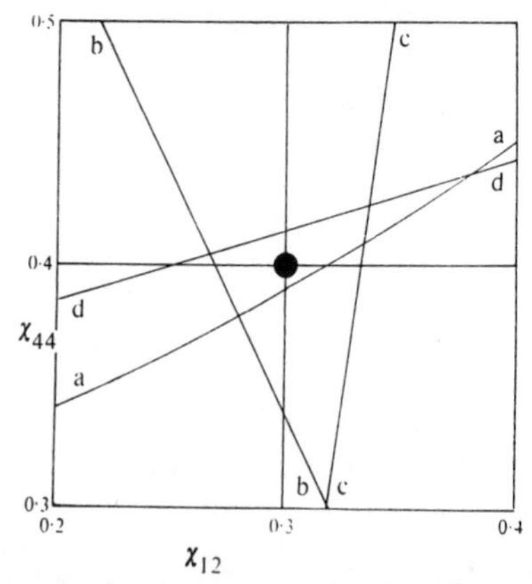

FIG. 2.15 Plot for diamond of corresponding values of χ_1 and χ_2 using the measured values of $\log_{10}(K[ABC]_{220}/K[110]_{220})$, as given below:

Line in diagram	*ABC*	$\log_{10}(K[ABC]_{220}/K[110]_{220})$
aa	001	0·27
bb	$1\bar{1}0$	0·45
cc	111	0·22
dd	001	0·41

(After Prince and Wooster, 1953).

2:6 The choice of diffuse reflections for elastic constant determination

One of the severe limitations on the applicability of diffuse X-ray reflections to the determination of elastic constants is the accuracy attainable in the measurements. In general a given measurement of diffuse intensity depends on several elastic constants but by a proper choice of reflecting planes and wave normals each measurement can be made to depend on only one or two elastic constants (Prasad and Wooster, 1955*c*).

This can be illustrated in the orthorhombic system. The elastic constants in this system are given by the following matrix:

$$\begin{bmatrix} c_{11} & c_{12} & c_{13} & 0 & 0 & 0 \\ & c_{22} & c_{23} & 0 & 0 & 0 \\ & & c_{33} & 0 & 0 & 0 \\ & & & c_{44} & 0 & 0 \\ & & & & c_{55} & 0 \\ & & & & & c_{66} \end{bmatrix}$$

The A_{ik} values obtained from this by equation (2.11) are:

$$A_{11} = c_{11}f_1^2 + c_{66}f_2^2 + c_{55}f_3^2$$
$$A_{22} = c_{66}f_1^2 + c_{22}f_2^2 + c_{44}f_3^2$$
$$A_{33} = c_{55}f_1^2 + c_{44}f_2^2 + c_{33}f_3^2$$
$$A_{12} = (c_{12} + c_{66})f_1f_2$$
$$A_{13} = (c_{13} + c_{55})f_3f_1$$
$$A_{23} = (c_{23} + c_{44})f_2f_3 .$$

For the wave normal direction [100] we have $f_1 = 1$, $f_2 = f_3 = 0$, and the $(A^{-1})_{ik}$ matrix becomes

$$\begin{bmatrix} \frac{1}{c_{11}} & 0 & 0 \\ & \frac{1}{c_{66}} & 0 \\ & & \frac{1}{c_{55}} \end{bmatrix}.$$

Then for relps having direction cosines g_1, g_2, g_3, we obtain from equation (2.34)

$$K[100]_{g_1g_2g_3} = \frac{g_1^2}{c_{11}} + \frac{g_2^2}{c_{66}} + \frac{g_3^2}{c_{55}} .$$

Thus when $g_1 = 1$, $g_2 = g_3 = 0$, the K-value depends only on c_{11}; when $g_1 = g_3 = 0$, $g_2 = 1$, the K-value depends only on c_{66}; and, lastly, when $g_1 = g_2 = 0$, $g_3 = 1$, the K-value depends only on c_{55}. It may not be possible or convenient to work with reflecting planes of such simple indices as $h00$, $0k0$, or $00l$. In this case recourse must be had to planes of the type $0kl$, $h0l$, $hk0$. The K-values may be expressed in terms of the elastic constants and the direction cosines of the normals to the reflecting planes. In Table 2.5, a few of the useful planes have been studied from this point of view.

The same table can be applied to the tetragonal system in classes of symmetry 42, $\bar{4}2m$, $4mm$, $4/mmm$, to the hexagonal system, and to the cubic system. The c_{ik} matrix has the same places filled with zeros in each of these systems and hence the K-values can be obtained by substituting the appropriate equalities. These equalities can at once be seen from the following c_{ik} matrices.

Tetragonal system classes 42, $\bar{4}2m$, $4mm$, $4/mmm$:

$$\begin{bmatrix} c_{11} & c_{12} & c_{13} & 0 & 0 & 0 \\ & c_{11} & c_{13} & 0 & 0 & 0 \\ & & c_{33} & 0 & 0 & 0 \\ & & & c_{44} & 0 & 0 \\ & & & & c_{44} & 0 \\ & & & & & c_{66} \end{bmatrix}.$$

TABLE 2.5

K-values in the orthorhombic system. Indices of planes hkl

Direction cosines of rekha $(f_1 f_2 f_3)$	$h00$	$0k0$	$00l$	$hk0$	$0kl$	$h0l$
100	$\frac{1}{c_{11}}$	$\frac{1}{c_{66}}$	$\frac{1}{c_{55}}$	$\frac{g_1^2}{c_{11}}+\frac{g_2^2}{c_{66}}$	$\frac{g_2^2}{c_{66}}+\frac{g_3^2}{c_{55}}$	$\frac{g_1^2}{c_{11}}+\frac{g_3^2}{c_{55}}$
010	$\frac{1}{c_{66}}$	$\frac{1}{c_{22}}$	$\frac{1}{c_{44}}$	$\frac{g_1^2}{c_{66}}+\frac{g_2^2}{c_{22}}$	$\frac{g_2^2}{c_{22}}+\frac{g_3^2}{c_{44}}$	$\frac{g_1^2}{c_{66}}+\frac{g_3^2}{c_{44}}$
001	$\frac{1}{c_{55}}$	$\frac{1}{c_{33}}$	$\frac{1}{c_{44}}$	$\frac{g_1^2}{c_{55}}+\frac{g_2^2}{c_{44}}$	$\frac{g_2^2}{c_{44}}+\frac{g_1^2}{c_{33}}$	$\frac{g_1^2}{c_{55}}+\frac{g_3^2}{c_{33}}$

Direction cosines of rekha $(f_1 f_2 f_3)$	$h00$	$0k0$	$00l$
$\frac{1}{\sqrt{2}}, \frac{1}{\sqrt{2}}, 0$	$\frac{2(c_{22}+c_{66})}{\{c_{11}(c_{22}+c_{66})+c_{22}c_{66}-2c_{12}c_{66}-c_{12}^2\}}$	$\frac{2(c_{11}+c_{66})}{\{c_{11}(c_{22}+c_{66})+c_{22}c_{66}-2c_{12}c_{66}-c_{12}^2\}}$	$\frac{2}{c_{44}+c_{55}}$
$0, \frac{1}{\sqrt{2}}, \frac{1}{\sqrt{2}}$	$\frac{2}{c_{55}+c_{66}}$	$\frac{2(c_{33}+c_{44})}{\{c_{22}(c_{33}+c_{44})+c_{33}c_{44}-2c_{23}c_{44}-c_{23}^2\}}$	$\frac{2(c_{22}+c_{44})}{\{c_{22}(c_{33}+c_{44})+c_{33}c_{44}-2c_{23}c_{44}-c_{23}^2\}}$
$\frac{1}{\sqrt{2}}, 0, \frac{1}{\sqrt{2}}$	$\frac{2(c_{33}+c_{55})}{\{c_{11}(c_{33}+c_{55})+c_{33}c_{55}-2c_{13}c_{55}-c_{13}^2\}}$	$\frac{2}{c_{44}+c_{66}}$	$\frac{2(c_{11}+c_{55})}{\{c_{11}(c_{33}+c_{55})+c_{33}c_{55}-2c_{13}c_{55}-c_{13}^2\}}$

Hexagonal system:

$$\begin{bmatrix} c_{11} & c_{12} & c_{13} & 0 & 0 & 0 \\ & c_{11} & c_{13} & 0 & 0 & 0 \\ & & c_{33} & 0 & 0 & 0 \\ & & & c_{44} & 0 & 0 \\ & & & & c_{44} & 0 \\ & & & & & \frac{1}{2}(c_{11}-c_{12}) \end{bmatrix}$$

Cubic system:

$$\begin{bmatrix} c_{11} & c_{12} & c_{12} & 0 & 0 & 0 \\ & c_{11} & c_{12} & 0 & 0 & 0 \\ & & c_{11} & 0 & 0 & 0 \\ & & & c_{44} & 0 & 0 \\ & & & & c_{44} & 0 \\ & & & & & c_{44} \end{bmatrix}.$$

The c_{ik} matrices for the remaining tetragonal classes, and for the monoclinic and triclinic systems of symmetry contain fewer terms which are zero. It is correspondingly more difficult to determine all the elastic constants present in the crystal. In principle it is possible to obtain all the necessary data by a combination of observations but up to the present no crystal of this lower symmetry has been studied by this means.

2:7 The variation with temperature of diffuse reflection

As the temperature of a crystal is raised the atoms vibrate with larger amplitudes. This affects the intensity of diffuse reflection in two opposite ways. The increase in the amplitude of atomic vibration corresponds to an increase in the amplitudes of the thermally excited elastic waves. This gives rise to increased intensity of the diffuse radiation. On the other hand, the increased atomic vibration diminishes the structure amplitude of the reflecting plane owing to the lessening of its truly planar character. These two effects thus operate in opposite directions. Further, the second effect is more pronounced the smaller the spacing between the planes. Thus the effect is more marked for higher order reflections than it is for the lower order reflections. In soft crystals with a low Debye temperature the diffuse intensity rises with increasing temperature up to a maximum, but with further increase in temperature it falls. The temperature at which this maximum occurs is known as the *inversion temperature* and is denoted T_h.

From equation (2.41) we have seen that the first order diffuse scattering power, D_1, is given by

$$D_1 \propto |F_T|^2 . T \tag{2.49}$$

for a given point in reciprocal space.

Now we may write

$$F_T = F_0 . e^{-M}, \tag{2.50}$$

where F_0 is the structure amplitude at absolute zero and M is the Debye-Waller temperature factor. The value of M is related to the amplitude u_s of atomic vibration in a lattice composed of one atom per unit cell according to the expression of James (1948, p. 193):

$$M = 8\pi^2 \overline{u_s^2} (\sin^2 \theta)/\lambda^2. \tag{2.51}$$

Combining equations (2.49) and (2.50), we have

$$D_1 \propto T.e^{-2M},$$

which when differentiated gives

$$dD_1/dT = e^{-2M}(-T.d(2M)/dT+1). \tag{2.52}$$

At the inversion temperature, T_h, the expression (2.52) becomes equal to zero. For a cubic crystal the value of T_h is derived by James (1948, p. 258) according to the equation:

$$T_h = \frac{mk.a^2.\Theta}{3h^2} \cdot \frac{1}{h^2+k^2+l^2},$$

where m, k, and h are the mass per unit cell, Boltzmann's constant, and Planck's constant respectively. The side of the unit cell is a and the Debye characteristic temperature is Θ. The inversion temperature decreases as $(h^2+k^2+l^2)$ increases and is thus not a constant for the crystal as a whole but only for one set of reflecting planes. This equation has been tested experimentally by Cartz (1955*a*), using single crystals of lead. In a further paper Cartz (1955*b*) related the thermal vibrations of atoms to the melting points. Canut and Amorós (1960) have discussed the same theory with special reference to molecular crystals.

III

THE SPECTRA OF ELASTIC VIBRATIONS

3:1 General considerations concerning elastic spectra in crystals

THE THERMAL motion of atoms in crystals is closely connected with the forces between neighbouring atoms. It is therefore important in studying the connection between crystal structures and their physical properties to know how atoms move relative to one another in crystals. A great help in this study is a knowledge of the distribution of energy associated with the thermally excited waves as a function of their frequency. Such a distribution curve is often described as an elastic spectrum. The distribution of the energy of vibration of the atoms among the possible frequencies and directions of travel of the elastic waves is a fundamental crystalline property which affects all the thermodynamics and elastic properties of the crystal. Various attempts were made to estimate the character of the elastic spectra but it was not until the development of the study of diffuse X-ray reflection was introduced that an experimental distribution curve of the elastic-wave frequencies could be obtained.

The first theory of the elastic spectrum is due to Debye (1912), who assumed that there was a maximum frequency for waves of all kinds, both longitudinal and transverse, and of the same magnitude for waves travelling in all directions. A curve for this spectrum can be drawn in the following way. As ordinate we plot $(dN^*/d\nu)/N$ and as abscissa ν. The quantity dN^* is the number of waves, including both longitudinal and transverse waves, having frequencies lying between ν and $\nu+d\nu$, and N is the total number of points on the elastic lattice. Both dN^* and N are counted over the volume of one reciprocal unit cell of the crystal lattice. The number N is very much greater than the number of waves which actually make a significant contribution to the elastic energy. The only practical use to which the number N is put is to act as a normalizing coefficient. In this particular application the area under the curve, Fig. 3.1, is equal to 3 because the total number of waves associated

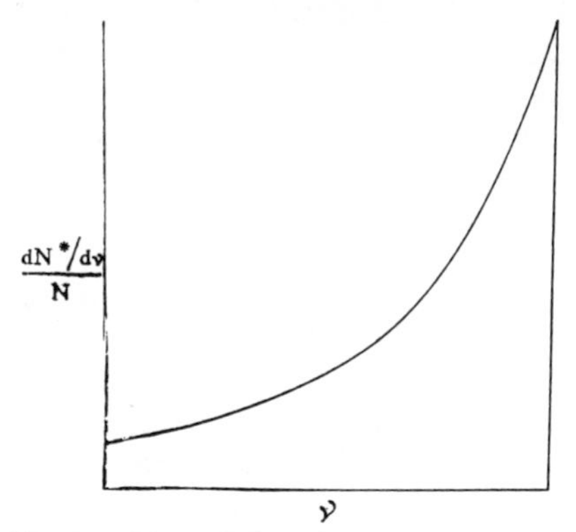

FIG. 3.1 The variation of the density distribution of frequencies with the frequency among elastic waves, according to Debye (1912).

with N points on the elastic lattice is $3N$, each point having one longitudinal and two transverse waves associated with it. All the subsequent curves showing $(dN^*/d\nu)/N$ as a function of ν have this area, no matter what may be the number of the peaks or the shape of the frequency spectrum. As we have seen in section 1:2.3, the number of points N is equal to the number of atoms in the crystal block.

3:2 Experimental determination of elastic spectra

Elastic spectra extend over a range of frequencies far greater than that involved in the determination of elastic constants described in Chapter II. Whereas elastic measurements seldom employ X-ray beams diffusely reflected in directions departing by more than 3° from the nearest Bragg reflection (using Cu $K\alpha$ radiation), the elastic spectra utilize the diffusely reflected radiation covering the whole angular range between successive Bragg reflections. Thus the intensities to be measured are smaller than those involved in elastic measurements and the contribution of Compton scattering and general background to the measured effects is correspondingly greater. Photographic methods have not been employed but only ionization and counter diffractometers. Olmer (1948) and Curien (1952*a*, *b*, *c*, *d*) have used an ionization chamber, and made special use of the possibility of adding a standard condenser to the electrometer system. This permits the determination of the ratio of the diffuse and direct X-ray beams in a simple and direct manner Cole and Warren (1952) compared the intensity of diffuse scattering from beta-brass with the intensity of scattering at high ϕ-angle from a thick block of paraffin wax. As described in section 1:3.2.7, the Compton scattering from such a block, consisting only of light elements, can be used to provide an absolute standard of intensity. Cole (1953) determined an approximate elastic spectrum for silver chloride by using a large single crystal. Walker (1956) re-determined the elastic spectrum of aluminium, and an approximate spectrum for beta-AuZn was found by Schwartz and Muldawer (1958).

The divergence corrections are important because a relatively large volume of reciprocal space is able to contribute its diffusely reflected beams at any given setting. The angles of divergence permitted by the apertures may be several degrees. In Olmer's experiments (1948) the correction factor was determined by dividing the domain in reciprocal space which contributed diffuse beams in any one setting into 48 parts. Each of these parts was considered as arising from a particular point on the focal spot of the X-ray tube. Tests using very fine slits and the incident X-ray beam were made to determine the contributions to the whole diffuse beam which could be made by each of the 48 elementary parts. Finally a global correction factor was evaluated appropriate to each point in reciprocal space. This factor varied between 0·7 and 1·2. Curien (1952*a*, *d*) followed a similar approach in determining the size and shape of the domains of divergence but was able to assume that for transverse waves the correction factor was always unity owing to the fact that the diffuse reflecting power varied but little over the divergence domain. For longitudinal waves Curien used a method of averaging over the domain based on the principle that by carrying out, on any given experimental curve, the same process of integration as is performed by the apparatus itself, a change is produced which is nearly equal to the correction required. He derived an expression giving the true value $g(x)$ as a function of the observed value $g^*(x)$ and its variation round about x up to the experimental limit of variation a, namely

$$g(x) = g^*(x) - \tfrac{1}{24}\{g^*(x+a) - 2g^*(x) + g^*(x-a)\}.$$

The method given in section 1:3.2.6, developed by Ramachandran and Wooster (1951*a*), can be applied to the measurements of elastic spectra as well as to the determination of elastic constants.

The correction necessary on account of Compton scattering is more important in the study of elastic spectra than in the determination of elastic constants. At points mid-way between relps the Compton scattering may amount to more than one-half of the total scattering. The Compton contribution is evaluated on the basis of theory and the expression often employed is that given in section 1:3.2.7 in connection with diamond. The expression for D_c, the Compton diffuse scattering power per unit cell, is given by the equation

$$D_c = P(Z - \sum f_{ec}^2)/B^3, \tag{3.1}$$

where P is the number of atoms (all of one sort) per unit cell, Z is the atomic number, f_{ec} is a function that describes the effect of the free atoms scattering incoherently, and B is the Breit-Dirac correction factor. The expression for B is

$$B = 1 + \frac{2h\lambda}{mc} \cdot \frac{\sin^2\theta}{\lambda^2}.$$

The values of $\sum f_{ec}{}^2$ are listed by Compton and Allison (1935, p. 782). The expression (3.1) refers to free atoms for which all recoil directions on receiving a quantum impact are possible. In crystals this condition is not fulfilled and an uncertainty arises in the calculation of the Compton scattering owing to this fact. Olmer (1948) found it necessary to reduce the calculated Compton scattering per electron in aluminium from about 0·55 to 0·50 for an angle of deviation of the X-rays through 30°, in order to obtain consistent results. As the Compton scattering may be more than half the total measured effect there is here the greatest source of uncertainty in the determination of the contribution due to the elastic waves. In the measurement of elastic constants the Compton effect can be neglected since it varies but slowly with the angle of deviation of the X-rays and merely forms part of the general background. However, in the study of elastic spectra and order-disorder phenomena the Compton effect plays a large part and must be evaluated in absolute measure. Curien (1958) has discussed methods of estimating the Compton contribution.

3:2.1 *The evaluation of the corrections due to second order thermal diffuse scattering*

The correction for second order thermal diffuse scattering may be large at distances removed from a relp by a large fraction of the distance between neighbouring relps. In one of Olmer's (1948) measurements the ratio of second to first order diffuse scattering power was 0·8. It is therefore necessary to make corrections more carefully for the effect of the X-rays twice reflected at acoustic waves than was necessary in the study of elastic constants.

Equation (2.43) is repeated for convenience:

$$I(\mathbf{QP}) = |F_T|^2 \frac{k^2T^2}{2\rho^2\tau^2}(OQ)^2(OP)^2 \sum_{i=1}^{3} \sum_{l=1}^{3} \frac{(e_{(i)k}g_k)^2}{(AQ)^2V_{(i)}^2} \cdot \frac{(e_{(l)m}g_m)^2}{(QP)^2V_{(l)}^2}. \tag{3.2}$$

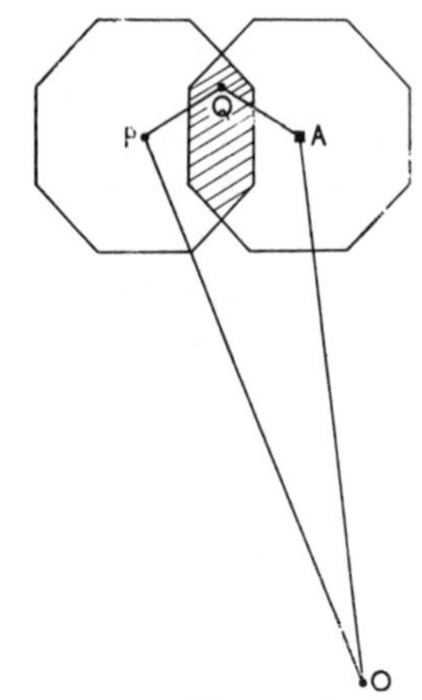

FIG. 3.2. Diagram showing the overlapping of Brillouin zones described about the relp A and the point P from which the scattering is being measured.

In this equation $I(\mathbf{QP})$ is the intensity of second order X-ray scattering associated with the wave having wave vector **QP** (Fig. 3.2). We may replace both **OQ** and **OP** by **R***. The expression for $I(\mathbf{QP})$ then contains the term

$$\sum_{i=1}^{3} \sum_{l=1}^{3} \frac{(e_{(i)k}g_k)^2 . (e_{(l)m}g_m)^2}{(AQ)^2(QP)^2}.$$

Now g_k, g_m are the direction cosines of the rel-vectors OP, OQ, which may be put equal to one another. Also, since the denominator of the above expression is a minimum when AQP is a straight line, a large part of the second order contribution is afforded by waves for which $e_{(i)k}$ and $e_{(l)m}$ are the same. We may, therefore, write with sufficient approximation

$$I(\mathbf{QP}) = |F_T|^2 \frac{k^2T^2}{2\rho^2\tau^2} R^{*4} \sum_{i=1}^{3} \frac{(e_{(i)k}g_k)^4}{V_{(i)}^4} \cdot \frac{1}{(AQ)^2(QP)^2}. \tag{3.3}$$

In section 1:2.3 we saw that the elastic waves which can cause first order diffuse scattering are represented by points within the first Brillouin zone, surrounding the chosen relp. Thus the point Q (Fig. 3.2) cannot be further from A than the boundaries of the first Brillouin zone constructed round A. Similarly, the points Q must lie within the Brillouin zone drawn round P. Thus if we represent the Brillouin zone by an octagon (Fig. 3.2) the shaded area common to both the zones drawn about A and P is the region within which the point Q must lie. Surrounding any point such as Q there must be eight relps of which A is the nearest. The zone drawn round P has a common volume with each of the eight zones corresponding to these eight corners of the reciprocal unit cell, and we must take this into account in the final estimation of the second order diffuse intensity.

Thus the contribution to the second order diffuse intensity, unlike that of the first order, is composed of contributions from many points of the elastic reciprocal lattice. For each point such as Q there is a term

$$\frac{1}{(AQ)^2(PQ)^2}$$

which determines its contribution to the whole effect. The sum of all these terms, each one corresponding to a particular position of Q, gives the total effect. Since the lattice points on the elastic lattice are so close together, we may replace this sum by an integral, I, and write

$$I = \frac{1}{\tau^*}\iiint \frac{d\tau^*}{(AQ)^2(PQ)^2}, \tag{3.4}$$

where τ^* is the volume of the reciprocal unit cell.

Up to this point all treatments of the subject are alike. The evaluation of the integral (3.4) is carried out with different approximations by different authors. We shall first consider the treatment given by Curien (1952*a*, *d*). The Brillouin zones are considered to be spheres concentric with the actual zones and having the same volume.

If we put $AP = 2y$ (following Curien's notation) and set the radius of the zone equal to R, then we obtain

$$I = \frac{\pi}{2R}\frac{1}{y/R}[I_1 - 2I_2],$$

where
$$I_1 = \int_0^{(4y/R)(1-y/R)} L(1-v)\frac{dv}{v}$$

and
$$I_2 = \int_0^{(R/y)-1} L\left(\frac{1-u}{1+u}\right)\frac{du}{u}.$$

Since y only occurs in the integration limits the integrals I_1 and I_2 were evaluated once for all and could then be applied to all cases. Numerical integration was employed, using values of the ratio y/R going from 0 to 1 in steps of 1/100. On inserting the integration limits, applicable to any particular relp and scattering reciprocal point, the value of I could be determined.

Ramachandran and Wooster (1951*a*) used an analytical method. The position of the point Q (Fig. 3.2) was determined by spherical polar coordinates (r, θ, ϕ). The distance QA is equal to r, the angle PAQ is θ, and the angle between the planes AOP and AQP is ϕ. A small element of volume in reciprocal space is then given by

$$d\tau^* = r^2 \sin\theta . dr . d\theta . d\phi.$$

The integration for I is performed (for the simplification of the analysis) between the limits

$$\begin{aligned} r&: \ 0 \text{ to } \infty \\ \theta&: \ 0 \text{ to } \pi \\ \phi&: \ 0 \text{ to } 2\pi. \end{aligned}$$

Then we may write for I,

$$I = \frac{1}{\tau^*} \int_0^\infty \int_0^\pi \int_0^{2\pi} \frac{1}{r^2} \cdot \frac{1}{r^2 - 2ry\cos\theta + y^2} \cdot r^2 \sin\theta . dr . d\theta . d\phi.$$

Since

$$\begin{aligned} \frac{1}{\sqrt{(r^2 - 2ry\cos\theta + y^2)}} &= \sum_{n=0}^{\infty} \frac{y^n}{r^{n+1}} p_n(\mu), \qquad r > y, \\ &= \sum_{n=0}^{\infty} \frac{r^n}{y^{n+1}} p_n(\mu), \qquad r < y, \end{aligned}$$

where $\mu = \cos\theta$ and $p_n(\mu)$ is the Legendre polynomial of order n (Jeffreys and Jeffreys, 1946), we may write

$$I = \frac{2\pi}{\tau^*} \int_0^y \int_{-1}^{+1} \left(\sum_{n=0}^{\infty} \frac{r^n}{y^{n+1}} p_n(\mu) \right)^2 d\mu \, dr + \frac{2\pi}{\tau^*} \int_y^\infty \int_{-1}^{+1} \left(\sum_{n=0}^{\infty} \frac{y^n}{r^{n+1}} p_n(\mu) \right)^2 d\mu \, dr.$$

Now
$$\int_{-1}^{+1} p_m(\mu) p_n(\mu) \, d\mu = \frac{2}{2n+1} \delta_{mn}$$

where
$$\begin{aligned} \delta_{mn} &= 1 \qquad \text{when } m = n \\ &= 0 \qquad \text{when } m \neq n. \end{aligned}$$

Hence

$$I = \frac{4\pi}{\tau^*} \left[\int_0^y \sum_{n=0}^{\infty} \frac{1}{(2n+1)} \frac{r^{2n}}{y^{2n+2}} \, dr + \int_y^\infty \sum_{n=0}^{\infty} \frac{1}{(2n+1)} \frac{y^{2n}}{r^{2n+2}} \, dr \right].$$

The two integrals within the square bracket are equal to one another and of value

$$\frac{1}{y} \sum_{n=0}^{\infty} \frac{1}{(2n+1)^2}.$$

We may therefore write

$$I = \frac{8\pi}{\tau^* y} \sum_{n=0}^{\infty} \frac{1}{(2n+1)^2}.$$

The limits of integration for r have been chosen as 0 to ∞ whereas, as we have seen above, only the volume of reciprocal space common to the two spheres described about A and P can contribute to the second order diffuse scattering associated with point A. If the upper limit of r is made the same as the radius of the first Brillouin zone and y is arbitrarily taken as $r/5$, then an estimate can be made of the error involved. The second term in I is now to be integrated between y and $5y$ instead of y and ∞. The effect of this is to reduce this term by about 20 per cent. The first term of the expression for I is unaffected and so the error in the estimate of the second order correction is about 10 per cent. Since the second order diffuse scattering correction is itself small, especially for points near to relps, this simplification in the analysis is justified. We may proceed further as follows:

$$\sum_{n=0}^{\infty} \frac{1}{(2n+1)^2} = \tfrac{3}{4} \sum_{n=1}^{\infty} \frac{1}{n^2} = \tfrac{3}{4}\zeta(2),$$

where $\zeta(x)$ is the Riemann ζ-function of x. Jahnke and Emde (1945) show that

$$\zeta(2) = \pi^2/6.$$

Thus, finally, the integral for I is given by

$$I = \pi^3/\tau^* y,$$

or

$$I = \pi^3/\tau^* K^*$$

in the notation adopted in this book.

We can now insert in equation (3.3) the value for the sum of all the terms $I(QP)$ which has been derived above and obtain for D_2, the second order diffuse scattering power per unit cell,

$$D_2 = \sum I(\mathbf{QP}) = |F_T|^2 \cdot \frac{k^2 T^2}{2\rho^2 \tau^2} \cdot R^{*4} \frac{\pi^3}{\tau^* K^*} \sum_{i=1}^{3} \frac{(e_{(i)k} g_k)^4}{V_{(i)}^4}$$

$$= \frac{\pi^3}{2} \cdot \frac{k^2 T^2}{\tau \rho^2} |F_T|^2 \frac{R^{*4}}{K^*} \sum_{i=1}^{3} \frac{(e_{(i)k} g_k)}{V_{(i)}^4}. \tag{3.5}$$

In the measurements of elastic constants the distance of a scattering point from the nearest relp is a small fraction of the side of the reciprocal unit cell. The value of $1/K^*$ ($1/AP$, Fig. 3.2) is therefore very small for all except the nearest relp and the sum $\sum 1/K^*$ taken over the eight relps at the corners of the reciprocal unit cell does not differ significantly from $1/K^*$ for the nearest relp. This is not the case in many of the measurements on elastic spectra, especially when the scattering point is almost mid-way between two relps. Equation (3.5) can be applied to such a scattering point if $1/K^*$ is replaced by $\sum_1^8 1/K^*$.

3:2.2 *Correction for third order diffuse scattering*

A small contribution to the scattering from any point of the elastic reciprocal lattice may be made by waves which have been scattered three times. In Fig. 3.3 scattering occurs from the elastic lattice point P as a result of scattering first from the elastic lattice points R and Q. Round the relp A both the first and the second Brillouin zones are drawn; around points R and P the first zone only is drawn in each case. The point Q must, as we saw in the section above, lie in the volume common to the first zones round R and A respectively. The point R must lie in the region common to the first zones round Q and P respectively. A more convenient and equivalent specification, however, is that R must lie in the region common to the second Brillouin zone drawn round A and the first zone round P, as shown by the shaded area round R in Fig. 3.3. The summation of the diffuse reflections which contribute to the scattering from any point P must be done over all the surrounding relps. There are 27, in the case of aluminium, which can satisfy the above geometrical requirements. The formulae are similar to those discussed in the previous section, and Olmer (1948) has given details of the method of numerical integration which can be employed. The magnitude of the correction is almost always less than the experimental errors.

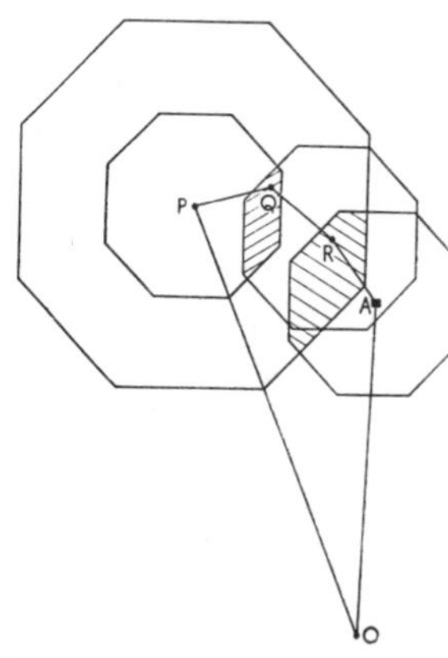

FIG. 3.3 Diagram showing the area common to (i) the first Brillouin zones described about the points A and R, and (ii) the second Brillouin zone described about A and the first zone about P.

3:3 **Results obtained**

The number of determinations of elastic spectra so far carried out using diffuse X-ray reflections is small and somewhat different problems have arisen in each of them. A summary is therefore given here of the various investigations in this field.

3:3.1 *Elastic frequency spectrum of aluminium*

The first complete investigation was that of Olmer (1948), who studied aluminium. Along a direction such as [001] only longitudinal waves can contribute to the diffuse scattering. A continuous series of measurements which included the relps 002, 004, 006 was made along this axis. Similarly, along the rekha stretching from 024 to $0\bar{2}4$ through 004 only transverse waves can contribute to the scattering. A continuous series of measurements between the relps 024 and $0\bar{2}4$ was made and the average of corresponding values determined. The values of the diffuse reflecting power, the various correction terms, and the calculated wave velocities for longitudinal waves are given in

Table 3.1. The corresponding values for transverse waves are given in Table 3.2. The values of H^* are given by $(2 \sin \phi/2)/\lambda$ (the radius of the reflecting sphere being taken as unity). The distance of the scattering point from the nearest relp is K^*; D is the diffuse scattering power per unit cell after correcting for all divergences; D_c, D_2, D_3 are the calculated Compton, second order, and third order thermal diffuse scattering powers respectively. D_1 is the first order diffuse scattering power, from which, after applying the quantum correction E/E_0, corresponding to the departure of the energy of the elastic quantum from kT, the velocity V is calculated. Finally, the figures in the column headed 'v' are obtained as the product of the corresponding figures in the columns headed K^* and V_l respectively. The corresponding data for the transverse waves are given in Table 3.2. Graphs showing V_l and V_t as a function of $K^*(\Lambda)$ are given in Figs. 3.4 and 3.5. The large change of the

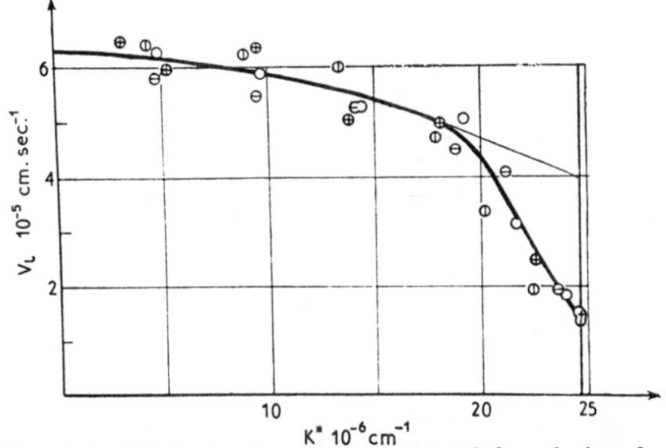

FIG. 3.4 Graph showing the dispersion of the velocity of longitudinal waves along the [100] axis in aluminium with the length of the wave vector. (After Olmer, 1948).

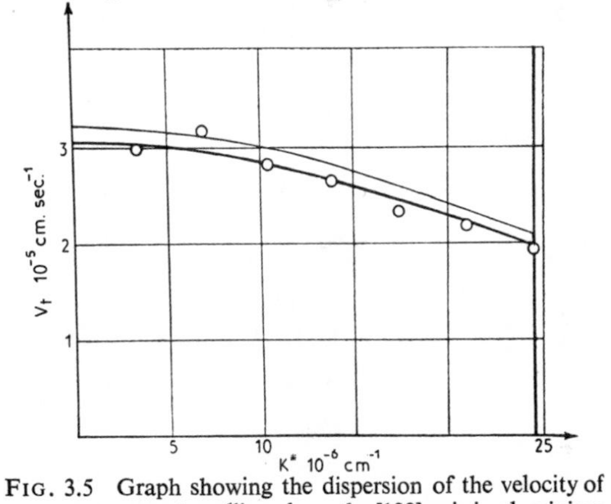

FIG. 3.5 Graph showing the dispersion of the velocity of transverse waves travelling along the [100] axis in aluminium with the length of the wave vector. (After Olmer, 1948).

TABLE 3.1

Data referring to scattering by longitudinal waves, travelling along [001]

H^* $\times 10^{-6}$	K^* $\times 10^{-6}$	*D measured corr. div.*	D_c	D_2	D_3	$D_c + D_2 + D_3$	D_1	E/E_0	V_l $\times 10^{-5}$	v_l $\times 10^{-11}$
Brillouin zone around 002										
49·38										
54·25	4·87	1·75	0·38	0·08	0·00	0·46	1·29	1·020	6·26	30·2
59·07	9·69	0·88	0·41	0·05	0·00	0·46	0·42	1·071	5·88	56·9
63·87	14·49	0·75	0·44	0·04	0·00	0·48	0·27	1·129	5·27	76·5
68·65	19·27	0·68	0·47	0·03	0·00	0·50	0·18	1·221	5·05	97·3
71·04	21·66	0·85	0·48	0·03	0·00	0·51	0·34	1·102	3·13	67·7
73·41	24·03	1·31	0·50	0·03	0·00	0·53	0·78	1·044	1·83	44·0
73·97	24·59	1·64	0·50	0·03	0·00	0·53	1·11	1·028	1·49	36·8
Brillouin zone around 004										
75·15	23·61	1·28	0·50	0·04	0·00	0·54	0·74	1·047	1·93	45·6
77·52	21·24	0·79	0·52	0·04	0·00	0·56	0·23	1·164	4·07	86·6
79·88	18·88	0·82	0·53	0·05	0·00	0·58	0·24	1·156	4·48	84·5
84·59	14·17	0·95	0·56	0·09	0·00	0·65	0·30	1·122	5·26	74·5
89·26	9·50	1·30	0·58	0·14	0·00	0·72	0·58	1·060	5·46	51·9
94·01	4·75	2·84	0·60	0·32	0·01	0·93	1·91	1·018	5·80	27·6

98·76					*Bragg reflection*					
103·10	4·34	2·80	0·64	0·36	0·01	1·01	1·79	1·017	6·40	27·8
107·70	8·94	1·30	0·65	0·19	0·01	0·85	0·45	1·069	6·24	55·7
112·20	13·44	1·02	0·67	0·12	0·01	0·80	0·22	1·143	6·00	80·6
116·70	17·94	0·97	0·68	0·09	0·01	0·78	0·19	1·158	4 71	84·7
118·95	20·19	1·02	0·69	0·08	0·01	0·78	0·24	1·102	3·36	67·8
121·20	22·44	1·41	0·70	0·07	0·01	0·78	0·63	1·042	1·91	43·0
123·40	24·64	1·70	0·71	0·07	0·01	0·79	0·91	1·026	1·41	34·7
					Brillouin zone around 006					
125·60	22·54	1·15	0·71	0·07	0·01	0·79	0·36	1·067	2·47	55·5
129·98	18·16	0·98	0·73	0·09	0·02	0·84	0·14	1·180	4·98	90·5
134·33	13·81	1·10	0·74	0·14	0·02	0·90	0·20	1·108	5·04	69·7
138·63	9·51	1·23	0·75	0·20	0·03	0·98	0·25	1·081	6·36	60·5
142·90	5·25	1·90	0·76	0·28	0·03	1·07	0·83	1·021	5·96	31·2
145·00	3·14	2·95	0·77	0·33	0·04	1·14	1·81	1·010	6·47	20·4
148·14					*Bragg reflection*					

TABLE 3.2

Data referring to scattering from points on either side of 004 *by transverse waves travelling along the line joining the reciprocal point* 004 *to* 024 *or* $0\bar{2}4$.

$H^* \times 10^{-6}$	$K^* \times 10^{-6}$	D *corr. div.*	D_c	D_2	D_3	$D_c+D_2+D_3$	D_1	E/E_0	$V_t \times 10^{-5}$	$v_t \times 10^{-11}$
98·76	0·00				*Bragg reflection*					
98·82	3·45	14·45	0·62	0·43	0·01	1·06	13·39	1·003	2·97	10·25
99·00	6·91	3·76	0·62	0·21	0·01	0·84	2·92	1·011	3·16	21·85
99·30	10·38	2·41	0·62	0·14	0·01	0·77	1·64	1·019	2·83	29·40
99·73	13·88	1·78	0·62	0·10	0·01	0·73	1·05	1·030	2·67	37·10
100·30	17·41	1·40	0·63	0·08	0·01	0·72	0·68	1·037	2·34	40·70
100·95	20·99	1·38	0·63	0·07	0·01	0·71	0·67	1·048	2·21	46·50
101·15	24·62	1·36	0·63	0·06	0·01	0·70	0·66	1·050	1·94	47·80

velocity with K^* for the longitudinal waves will be noticed. The rapid fall in the longitudinal velocity occurs when the wavelength falls below about 5A ($K^* = 20 \times 10^6$ cm^{-1}). When K^* is zero the velocity should correspond with that determined by static and the usual dynamic methods. The value of the velocity for infinitely long elastic wavelengths derived from the work of

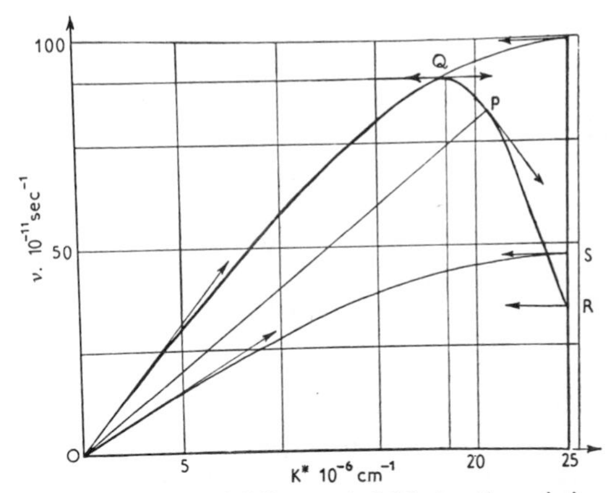

FIG. 3.6 The upper full-line graph OQR gives the variation of frequency with wave vector for longitudinal waves along [100] and the lower graph the corresponding variation for transverse waves travelling along [100]. (After Olmer, 1948).

Goens (1933) is $6{\cdot}30 \times 10^5$ cm/sec and the curve of Fig. 3.4 has been made to pass through this point. The results for the transverse waves obtained by the classical and X-ray methods do not quite agree, the value due to Goens (1933) being $3{\cdot}25 \times 10^5$ cm/sec whereas that derived from the X-ray measurements is $3{\cdot}06 \times 10^5$ cm/sec.

3:3.1.1 *Dispersion of the frequencies of longitudinal and transverse waves*

When the values of ν_l and ν_t given in Tables 3.1 and 3.2 are plotted against K^* we obtain the curves shown in Fig. 3.6. There are certain simple geometrical properties of these curves. Since $\nu = V.K^*$ the slope of any line such as OP gives the velocity corresponding to the point P. Further, since

$$K^* = 1/\Lambda, \qquad dK^* = -\frac{1}{\Lambda^2}d\Lambda.$$

Hence
$$\frac{dV}{dK^*} = -\Lambda^2 \frac{dV}{d\Lambda}$$

and
$$K^* \frac{dV}{dK^*} = -\Lambda \frac{dV}{d\Lambda}.$$

Now the group velocity of the wave, U, for wavelengths near to Λ is given by

$$U = V - \Lambda \frac{dV}{d\Lambda} = V + K^* \frac{dV}{dK^*} = \frac{d}{dK^*}(V.K^*)$$

$$= d\nu/dK^*, \tag{3.6}$$

i.e. U is given by the tangent to the curves of Fig. 3.6. Thus at Q the group velocity becomes zero and between Q and R negative. The group velocity of the transverse wave becomes zero at the limit $K^* = 24 \times 10^6$ set by the boundary of the first Brillouin zone. The consequence of a zero value of the group velocity at a given value of Λ is that at a given point in the crystal the atomic displacements due to waves of wavelength near to Λ are constant, i.e. the atoms do not vibrate and some atoms remain undisplaced. This, of course, only applies to the effects of waves of wavelength near to the particular value Λ for which the tangent to the ν-curve is horizontal.

3:3.1.2 *Calculation of the frequency spectrum*

Olmer did not make measurements in directions other than those quoted above but he assumed that it would be legitimate to neglect differences between the axis [100] and other directions in deriving the frequency spectrum. The number of points per unit reciprocal cell on the elastic lattice is N, where N is the total number of atoms in the crystal (see section 3:1). If τ^* is the volume of the unit cell of the reciprocal lattice and we suppose the elastic lattice points to be uniformly distributed, then we may write for the density of points,

$$\sigma = N/\tau^* = N.\tau,$$

where τ is the volume of the Bravais unit cell.
The number of wave vectors of lengths lying between $|K^*|$ and $|K^*+dK^*|$, corresponding to a range of frequency $d\nu$, is dN^*.

To each wave vector there corresponds one longitudinal and two transverse waves and we may write

$$dN_l^* = 4\pi\sigma K^{*2}\, dK^*$$

$$dN_t^* = 8\pi\sigma K^{*2}\, dK^*$$

for the numbers of longitudinal and transverse waves respectively having wave vectors lying between $|K^*|$ and $|K^*+dK^*|$.

We have seen from equation (3.6) that

$$dK^* = \frac{d\nu}{U} \quad \text{and} \quad K^* = \frac{\nu}{V}.$$

Hence $$dN_l^* = 4\pi\sigma \frac{\nu^2}{U_l V_l^2}\, d\nu, \qquad dN_t^* = 8\pi\sigma \frac{\nu^2}{U_t V_t^2}\, d\nu,$$

and finally

$$dN^* = dN_l^* + dN_t^* = N.\tau 4\pi \left[\frac{1}{U_l V_l^2} + \frac{2}{U_t V_t^2} \right] \nu^2 \, d\nu.$$

The quantity in the square bracket can be evaluated for each value of ν from the curves in Figs. 3.4 and 3.5. Thus $(dN^*/d\nu)/N$ can be evaluated over the whole observed frequency range. This calculation results in Fig. 3.7. The sharp peaks are associated with points on the curves of Fig. 3.6 where U becomes zero. As ν increases, the first peak occurs where the tangent to the ν_l curve of Fig. 3.6 becomes horizontal at R; the second occurs at the point where the ν_t curve crosses the vertical line corresponding to the maximum frequency at point S, and the third peak occurs at a point corresponding to point Q on the ν_l curve. It will be seen that this frequency distribution curve is different from the original form put forward by Debye, which had only a single peak, though it is rather similar to subsequent proposed forms, which allowed for transverse as well as longitudinal waves. The integral of the area under the curve of Fig. 3.7 should be equal to 3 since the total number of points on the elastic reciprocal lattice is $3N$ per unit cell of the reciprocal lattice. In fact, for the curve obtained by Olmer it is 2·86 and the error arises in integrating numerically over the peaks of the curve.

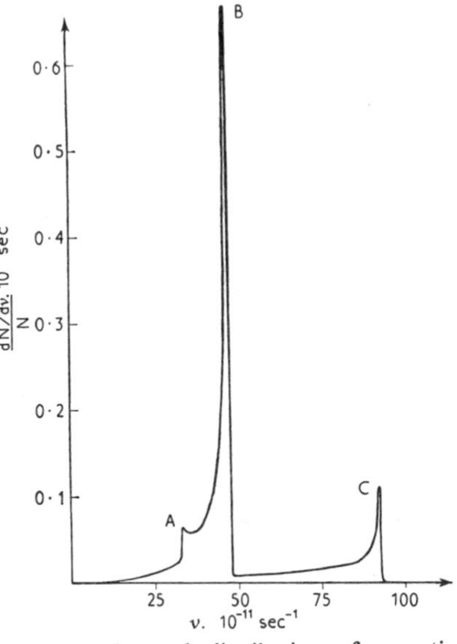

FIG. 3.7 Spectral distribution of acoustic frequencies in aluminium for waves travelling along [100]. The peak A corresponds to the maximum value of K^* for the longitudinal waves (point R, Fig. 3.6). The peak B corresponds to the maximum K^* for transverse waves (point S, Fig. 3.6), and point C corresponds to the maximum frequency for longitudinal waves (point Q, Fig. 3.6). (After Olmer, 1948).

Olmer's work has been described in detail because it brings out clearly the principles involved. The study of aluminium has been repeated by Walker (1956), using more modern methods. One difference between the results obtained is that Walker's ν/K^* curve for longitudinal waves along [100] does not reach a maximum but rises steadily, becoming horizontal at the boundary of the Brillouin zone. This difference is attributed to the presence of a small amount of $\lambda/2$ in the incident radiation used by Olmer. The final spectral distribution obtained by Walker is rather different from that of Fig. 3.7. The peaks are lower relative to the general background and the peak corresponding to B in Fig. 3.7 is lower than that corresponding to C.

3:3.2 *Elastic frequency spectrum of alpha-iron*

Curien (1952*a*, *b*, *c*, *d*) studied the elastic properties of iron by essentially the same technique as that used by Olmer. The nature of the material introduced slight differences in treatment. The radiation used was Mo $K\alpha$, monochromatized by reflection at a curved quartz crystal. A filter of aluminium 0·2 mm thick was placed before the ionization chamber to reduce the intensity of the fluorescent radiation. The diffuse reflection due to longitudinal waves and also that due to transverse waves travelling along the tetrad, triad, and diad axes was studied. Crystals were cut in different orientations with respect to the crystallographic axes to make this extensive exploration possible. Some of the curves showing the variation of frequency against wave vector, K^*, are similar in shape to those obtained for longitudinal waves travelling parallel to [100] in aluminium (Fig. 3.6). This is true for longitudinal wave travelling parallel to the axes [100] and [111], while for longitudinal waves parallel to [110] the variation is similar to that of the transverse waves shown in Fig. 3.6. The curve for transverse waves travelling perpendicular to the axis [100] is also like that of the transverse waves in Fig. 3.6. The corresponding curve for transverse waves travelling perpendicular to [111] is almost a straight line. A quantity $n_\nu\, d\nu$, defined according to the equation

$$n_\nu\, d\nu = \frac{1}{\tau^*} K^{*2}\left(\frac{dK^*}{d\nu}\right) d\nu, \tag{3.7}$$

is proportional to the number of waves travelling in a given direction and having frequencies lying between ν and $\nu + d\nu$ (previously defined as dN^*). The quantity $n_\nu\, d\nu$ can be evaluated from the curves giving ν as a function of K^*. The direction of a given wave normal is now specified by two angles θ and ϕ and, using spherical harmonics, we may write

$$n(\nu, \theta, \phi) = \sum_i n_i(\nu) Y_i(\theta, \phi), \tag{3.8}$$

where $n(\nu, \theta, \phi)$ is now the value of n_ν along a particular direction defined by the angles θ and ϕ, $n_i(\nu)$ is a quantity depending only on the frequency, and $Y_i(\theta. \phi)$ is the usual Legendre function. If the function $n(\nu, \theta, \phi)$ is evaluated for a number, R, of known directions, then we obtain the same number of equations, similar to (3.8). If only first terms in the $Y_i(\theta, \phi)$ factor are used, the R linear equations thereby obtained can be solved. The values of θ, ϕ, and n for the tetrad, triad, and diad axes are given in Table 3.3, where n_1, n_2, and n_3 are evaluated from the experimental curves according to equation (3.8). The spherical harmonics which come into account are those which are in-

TABLE 3.3
Corresponding values of θ, ϕ, and n

Direction	θ	ϕ	n
Tetrad axis	0	0	n_1
Diad axis	$\pi/4$	0	n_2
Triad axis	$\cos^{-1} 1/\sqrt{3}$	$\pi/4$	n_3

variant in the transformations corresponding to cubic symmetry. Here the three first cubic harmonics have been used in evaluating

$$\frac{dN^*/d\nu}{N} = \iint n(\nu, \theta, \phi) \sin\theta \, d\theta \, d\phi, \tag{3.9}$$

which, on inserting the actual quantities, becomes

$$\frac{dN^*/d\nu}{N} = \frac{4\pi}{469}[162n_1 + 208n_2 + 99n_3].$$

Finally, after weighting in this way the contributions to the total number of elastic reciprocal points due to the waves travelling in various directions, the final distribution curve is obtained as shown in Fig. 3.8. Its area has been

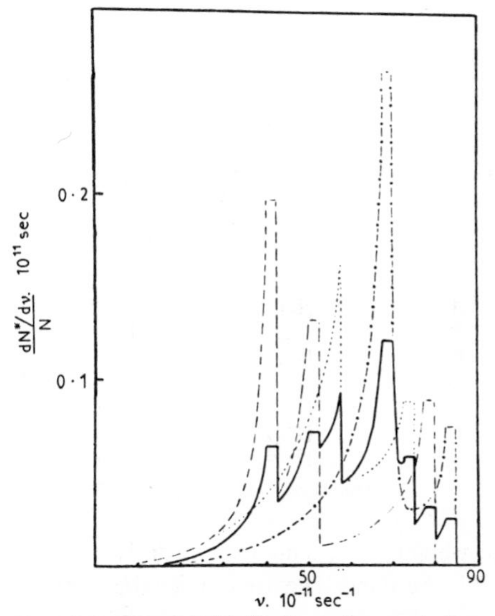

FIG. 3.8 Spectral distribution of acoustic frequencies in alpha-iron for waves travelling in directions [100] (dash-dot curve), [110] (dash curve), and [111] (dotted curve). The full-line curve is representative of frequencies of waves travelling in all directions. (After Curien, 1952*d*).

made equal to 3 for reasons explained in section 3:3.1. The contributions to this composite curve due to the waves having wave normals directed along the tetrad, diad, and triad axes are shown separately in Fig. 3.8.

3:3.3 *Elastic frequency spectrum of β-brass*

Cole and Warren (1952) applied the same method to beta-brass which, like alpha-iron, is body-centred cubic. The only difference in the treatment was the weighting of the contributions due to the waves associated with the wave normal directions [100], [110], and [111] respectively.

The factors used by Curien, namely, 162/469 = 0·35, 208/469 = 0·44, and 99/469 = 0·21, differ somewhat from those used by Cole and Warren, namely, 6/26 = 0·23, 12/26 = 0·46, and 8/26 = 0·31, respectively. Apart from

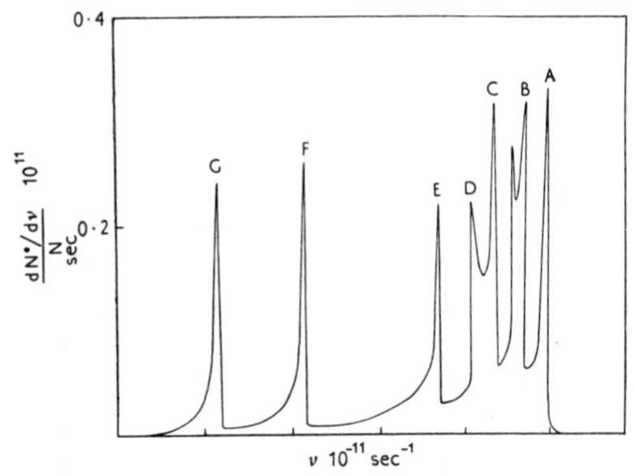

FIG. 3.9 An approximate frequency spectrum of elastic waves in beta-brass at 295°K. The peaks correspond to horizontal tangents to the ν/K^* curves and are described as follows:

	Wave normal direction	*Mode*
A	[100]	transverse
B	[111]	longitudinal
C	[110]	longitudinal
D	[100]	longitudinal
E	[110]	transverse (1)
F	[111]	transverse
G	[110]	transverse (2)

(After Cole and Warren, 1952).

this difference in weighting, the manner of deriving the frequency spectrum shown in Fig. 3.9 is the same. The area under this experimentally determined curve is 2·7, instead of 3·0 as it should be according to the theory.

The assignation of the peaks to various modes of vibration is given in Fig. 3.9. Cole and Warren are not sure that the peaks *F* and *G* are well established and suggest that there may be a series of smaller peaks between the positions of *F* and *G* instead of the two large peaks.

IV

DIFFUSE X-RAY SCATTERING DUE TO STRUCTURAL IMPERFECTIONS

4:1 Historical introduction

THE DEVELOPMENT of this part of the study of diffuse reflections now extends over more than three decades. The first observation was due to Mauguin (1928), who investigated the diffuse layer lines observed on rotation photographs of biotite mica. Over this period certain themes have been constantly recurring. The layer structures have been popular: one group of investigators has extended Mauguin's work on micas—Hendricks (1938, 1939, 1940), Hendricks and Jefferson (1938, 1939), Hendricks and Teller (1942), Méring (1949), Brindley, Oughton, and Robinson (1950). The nature of the faults in the stacking of the successive atomic layers in this type of structure is satisfactorily understood. Another type of layered structure, exemplified by CdI_2 and metallic cobalt, in which there is an alternation of cubic and hexagonal close-packing, has received much attention. The first paper in this group is due to Shôji (1933), who investigated the partial transformation of zinc blende (ZnS) into wurtzite. Laue photographs of partially transformed material showed continuous diffuse zone circles as well as sharp spots. Bijvoet and Nieuwenkamp (1933) studied the mixed structure of $CdBr_2$, which has an irregular packing of the layers parallel to (00·1). Ketelaar (1934) studied $NiBr_2$, which is similar to $CdBr_2$ in this structural feature. The investigation of the alternating cubic and hexagonal close-packing in cobalt was begun by Edwards and Lipson (1942). Sharp and diffuse reflections occur on the same photograph and this was given a theoretical explanation both by Edwards and Lipson (1942) and also by Wilson (1942). At this time Bragg and Lipson (1943) introduced the optical diffractometer which has proved so useful in the study of this subject. Zachariasen (1947) gave a mathematical analysis which afforded a direct determination of stacking disorder in layer structures. Jagodzinski (1949*a*, *b*, *c*) analyzed theoretically the diffuse intensity distributions which he observed on a number of photographs of various inorganic crystals. The diffuse scattering of X-rays by plastically deformed single crystals of aluminium was investigated by Kolontsova (1950), who showed that the thermal diffuse scattering remains the most important diffuse effect until 4 to 8 per cent. of plastic deformation is achieved. Paterson (1952) developed a theory applying to a face-centred cubic crystal with faults in the stacking of the atomic planes. This theoretical work was extended by Hosemann and Bagchi (1954). Gevers (1954*a*, *b*) also gave a general theory of the effects due to stacking faults and transformations in close-packed structures. The now classical example of sharp and diffuse spots occurring together in Wollastonite was published by Jeffrey in 1953. The stacking faults which occur in ZnS crystals were studied by Krumbiegel and Jost (1955). Chayes (1956) applied the optical diffractometer to the study of the diffuse patterns given by disordered layer structures. Dornberger-Schiff (1956) introduced a systematic treatment of order-disorder structures, having special reference to those consisting of perfectly regular layers stacked in irregular ways. This treatment was applied to the study of the structure of Purpurogallin (1957). Longuet-Escard and Méring (1957) dealt with the laws of stacking which apply in the structure of nickel hydroxide. Hoshino (1957) investigated the structure and diffuse effects given by alpha-AgI. Studying the sharp and diffuse reflections afforded by members of the plagioclase felspars, Bown and Gay (1958) traced the unusual distributions of these diffuse reflections in reciprocal space. A theoretical study of the influence of stacking faults in face-centred cubic alloys was given by Willis (1959).

The second theme which frequently recurs in the papers on diffuse reflections is the structure of diamond. A brief summary of the opposing viewpoints taken in this classical controversy has been given in Chapter I. Only certain developments need be mentioned here. Rules relating the intensities of the various diffuse spikes associated with particular reflections were put forward by Hoerni and Wooster (1952*a*, 1955). It was suggested by Frank (1956) that silicon atoms segregated in planes parallel to $\{100\}$ were responsible for the spikes. This proposal was not supported by quantitative work on measurements of the intensities of spikes (Caticha-Ellis and Cochran, 1957). The observation that nitrogen is an impurity in diamond has led Elliott (1960) to propose a model involving nitrogen atoms segregated into planes parallel to $\{100\}$ and this seems the most promising theory so far advanced.

A third subject which has claimed much attention has been the diffuse scattering shown by the age-hardening alloys. Preston (1938*a*, *b*) and Guinier (1938) independently published the first papers on the diffuse effects seen in Al-4 per cent. Cu alloys and interpreted the observations as being due to the segregation of copper atoms in thin platelets arranged parallel to the $\{100\}$ planes. Guinier (1942, 1943*a*, *b*, 1944) applied the method to several of the age-hardening alloys, such as Al-Ag, Cu-Be, Al-Mg-Si. He gave a review of his experimental and theoretical approach in 1945. An interpretation was developed according to which the precipitated zones not only increase in size as precipitation proceeds but also change their structure and composition (Guinier, 1950). Pashalov (1950) also studied some of the age-hardening alloys, particularly the alloys of 6·5 per cent. Cu-Ag. Bagaryatskii (1951) presented a theoretical discussion of the diffuse scattering from age-hardening alloys. Newkirk *et al.* (1951) investigated the aging process in an alloy of CoPt. Another type of inhomogeneous structure was described by Hargreaves

(1951), who dealt with the Cu-Ni-Fe alloys. He showed that the diffuse scattering is due to modulated structures in which two tetragonal phases occur. Guinier (1952) carried the interpretation of the process of age-hardening a stage further and concluded that irregular regions arise, within which segregation of one sort of atom occurs, and that the atomic arrangement within these regions departs more and more from that of the matrix as age-hardening proceeds. Finally, a crystal structure arises which differs from that of the matrix. Walker (1952) studied the diffuse scattering from Cu-Pt alloys between 25°C and 890°C and found that the segregated atoms tend to occur in small platelets. Bagaryatskii (1951, 1952) repeated the work on the aging of certain aluminium alloys. In particular he studied 3 per cent. Cu-1·15 per cent. Mg-Al and 4 per cent. Cu-Al. Aging at room temperature was compared with that at 170°C and 218°C. Murakami (1953*a*, *b*) presented a theory of the diffuse scattering due to binary alloys and considered both the irregular distribution of foreign atoms and also the segregation of atoms in platelets and other forms. Toman (1955, 1957) presented a theoretical analysis of the Al-4 per cent. Cu diffuse streaks. He calculated the concentration of copper atoms in successive atomic planes of Guinier-Preston zones and found that the platelets were thicker than previous investigators had found. Even at a distance of five atomic diameters from the central plane of the platelet the copper atoms still accounted for 17 per cent. of the composition. Doi (1957, 1960) also considered the same problem but came to the ocnclusion that the zones are not more than three atoms thick when the diffuse streaks are produced. An analogue of the age-hardening alloys is afforded by solid solutions of NaCl-$CaCl_2$ (Suzuki, 1958) from which precipitation occurs in platelets.

A similar problem to that of the age-hardening alloys is presented by the alloy Cu_3Au. It was in connection with this alloy that the classical conceptions of order and disorder were developed. The early work (Sykes and Jones, 1936; Jones and Sykes, 1938) was done on powder photographs and only the line broadening could be measured. Work on single crystals was more informative. Guinier (1945) and MacGillavry and Strijk (1946*a*, *b*, *c*) studied single crystals of Cu_3Au. The interpretation that is now accepted was first obtained from the experimental work of Edmunds, Hinde, and Lipson (1947) and the theoretical analysis of Wilson (1947). It was shown that in reciprocal space there are diffuse disks which, according to their indices, are set perpendicular to the x-, y-, z-axes respectively. A fuller account of the theory was given by Wilson (1949*a*). This pattern in reciprocal space corresponds to a particular ordering of atoms of copper and gold relative to one another in the actual crystal structure. Cowley (1950*b*) described an approximate theory of long and short range order in alloys, including Cu_3Au. He also made measurements of the degree of order (1950*a*). Taylor, Hinde, and Lipson (1951) applied the optical diffractometer to the study of this ordering process. Murakami (1953*a*, *b*) gave a theory of the diffuse scattering due to binary alloys, which included the example of Cu_3Au, and he also found a similar distribution in reciprocal space to that of earlier workers. Borie and Warren (1956) studied the changes in Cu_3Au near 600°C by means of diffuse reflection. Chipman (1956) showed that below the critical temperature there is a marked short range order among the wrongly occupied sites. Herbstein, Borie, and Averbach (1956) measured the intensities of Bragg reflections in various binary alloys, including Cu_3Au, and correlated this work with the diffuse reflections. Hirabayashi and Ogawa (1956) investigated the ordering process in single crystals of CuAu and related the changes in the X-ray pattern to the kinetics of the ordering process. Batterman (1957) made a study of the degree of order in the corresponding alloy of composition $CuAu_3$; the coefficients of short range order were found to be similar to those of Cu_3Au. A study of the order in the alloy $CdMg_3$ was carried out photographically by Steeple and Edmunds (1956).

In addition to the special study of the alloy Cu_3Au in its various states of atomic ordering, there have been numerous studies of a more general character. Most of this work has been concerned with the variation of the intensity across powder lines. In some materials it is impossible to obtain single crystals of a sufficient size but in a number of studies the powder line has been used, mainly because of experimental simplicity of the observations, rather than because of the nature of the crystal. The simplicity of experimentation is not matched by the simplicity of theoretical interpretation, as may be seen from the many attempts at solving the problem. Landau (1936) derived formulae for the powder-line profile associated with diffuse effects produced by lamellar structures. Huang (1947), in a much quoted theoretical paper, discussed both normal and diffuse reflections to be expected from solid solutions. Using a single crystal, Chipman and Warren (1950) applied diffuse measurements to the study of a long range order in beta-brass. Though most work has been directed to discovering the diffuse effects around and between the normal relps, some research has been applied to low angle X-ray scattering which gives information concerning the region close to the origin of the reciprocal lattice. An example of this kind of study is the paper by Flinn and Averbach (1951) on the solid solutions of gold and nickel. Kakinoki and Komura (1952) derived theoretically the intensity of X-ray diffraction by a crystal having one-dimensional disorder. Kakinoki and Komura (1954*a*, *b*) continued this type of approach and applied the results to a close-packed structure. Flinn, Averbach, and Rudman (1954) interpreted the diffuse powder patterns of solid solutions of Al-Zn at 400°C, and also those of Ni-Au. Cochran (1956) applied the concept of the Fourier transform to the problem of structural defects and showed that the intensity of X-ray scattering depends on the Fourier transforms of the defects taken separately. Cochran and Kartha (1956*a*, *b*) applied this theory to several problems including scattering by interstitial atoms. The anomalous diffuse scattering of white tin was described, but not accounted for, by Prasad and Wooster (1956*a*). Wagner (1957*a*) studied experimentally the effect of filing alpha-brass, at −160°C and at room temperature, on the stacking faults produced. Similar investigations were also carried out on silver and aluminium (Wagner, 1957*b*). Borie (1957) developed a theory of the diffuse scattering from a binary alloy. Suoninen and Warren (1958) determined the short-range order in beta-Ag-Zn alloys. Houska and Averbach (1959) studied the diffuse effects which may be observed in powder photographs due to structural disorder in cubic solid solutions. Chipman and Paskin (1959*a*, *b*) went over familiar ground more thoroughly and compared the experimental results in powder photographs of copper and lead with various theoretical approximations. A new feature of this work was the determination of the correction to the observed integrated intensity of the Bragg reflection due to the temperature diffuse scattering.

A number of papers have been written on the structural disorder in molecular crystals. Chapter V is devoted to diffuse effects in molecular crystals and the following papers are mentioned only to complete this brief account of the study of structural disorder. Powell and Huse (1943) formed crystals of picryl halogenides containing hexamethylbenzene. These are layer structures showing disorder. James and Saunder (1947, 1948) investigated crystals containing molecules of the type 4:4′-dinitrodiphenyl with 4-halogen diphenyl, which also show disorder. Kasper, Lucht, and Harker (1950) determined the structure of decaborane, $B_{10}H_{14}$. This crystal has a structure which readily leads to a type of twinning, not on the macro scale as is very common in minerals, but on the scale of a few molecular diameters. Associated with this are special arrangements of sharp and diffuse reflections. Deas (1952) considered theoretically the diffuse effects to be expected from a random assemblage of molecules having a partial alignment.

Dislocations give rise to diffuse scattering and the study of this has received a small amount of attention. Wilson (1949*b*) determined the reciprocal equivalent of a screw dislocation. Except for reflecting planes which are parallel to the axis of the dislocation the intensity of reflection at the centre of any relp is zero. The scattering which in ideal lattices is concentrated into point-like relps becomes, in the lattice distorted by a screw dislocation, spread out into disk-like rings surrounding the relps. The plane of the disk is perpendicular to the axis of the screw dislocation. Wilson (1952, 1955) applied this study to the breadth of powder lines. Willis (1957*a*, *b*) used the optical diffractometer in order to study the diffuse effects associated with dislocations; his published photographs refer to edge dislocations.

4:2 **Diffuse scattering from layer structures**

Diffuse scattering may arise because of structural departures from an ideal lattice in one, two or three dimensions. If the structure is composed of identical, parallel planes of atoms which are displaced parallel to one another in a regular or irregular manner from the ideal positions of a perfect lattice, then the irregularity is one-dimensional. If the structure consists of parallel chains of atoms which are displaced sideways relative to one another from their ideal positions, then the departure from structural regularity is two-dimensional. Lastly, if by the insertion of separate foreign atoms, or groups of atoms, the structure is locally swollen or contracted, then the irregularity of the structure is three-dimensional. Disorder of the atomic arrangement may be associated with effectively infinite planes or with chains of atoms or with localized regions. There are, therefore, an almost infinite variety of possible structural defects which can give rise to diffuse effects.

4:2.1 *Sharp and diffuse reflections*

A striking feature of many photographs obtained from crystals having certain types of structural disorder is the occurrence, side by side, of *sharp* and *diffuse* spots. We shall trace the manner in which these can arise from the structural disorder, using optical analogues to illustrate the phenomena.

4:2.1.1 *Wollastonite*

The simplest of the types of one-dimensional disorder is associated with the mineral Wollastonite, $CaSiO_3$. This was studied by Jeffrey (1953), who showed that on a Weissenberg photograph of a crystal from Devon, oscillating about the b-axis, the even-order layer line photographs had sharp spots whereas the spots on the odd-order layer line photographs were diffuse and drawn out into streaks along curves corresponding to a^*. Willis (1958) used an optical diffractometer to investigate this type of structure. To concentrate on the essentials he chose a simple cubic lattice rather than the monoclinic lattice of Wollastonite. The type of displacement of the horizontal rows of atoms is shown in Fig. 4.1*a*. The optical masks carried many thousands of

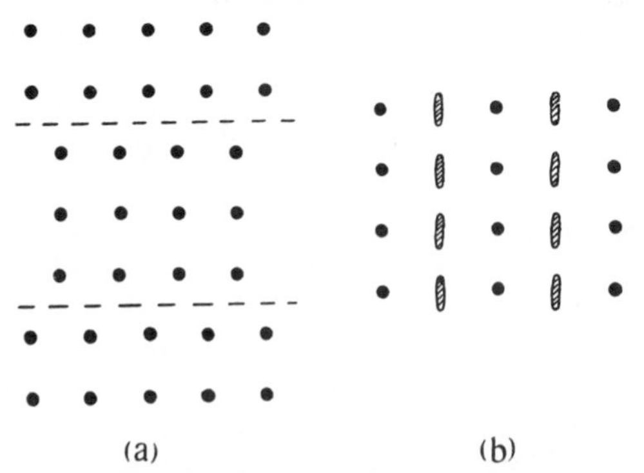

FIG. 4.1*a* Diagram representing an idealized Wollastonite-type of lattice faulting.

FIG. 4.1*b* Diagram of the optical diffraction pattern corresponding to the mask of which Fig. 4.1*a* is a small part.

holes and Fig. 4.1*a* shows only a small portion of such a mask. The dotted lines show where a displacement of half a cell side has occurred. Fig. 4.1*b* gives the corresponding diffraction pattern produced by such a mask. It will be seen that some spots are sharp while others are diffuse and the latter are drawn out in a direction normal to the dotted lines of Fig. 4.1*a*. In the direction of the dotted lines every row of points is quite regular. The diffraction pattern produced by such a single row of points would be a set of vertical continuous lines passing through the sharp as well as the diffuse spots of Fig. 4.1*b*. If the arrangement of points were everywhere like that between the dotted lines of Fig. 4.1*a* then the diffraction pattern would consist of sharp spots only, occurring in the positions of both the sharp and the diffuse spots in Fig. 4.1*b*. At this point it is convenient to use the concept of the *stacking lattice*—Dornberger-Schiff (1956). In Fig. 4.2 the same points have been drawn as occur in Fig. 4.1*a* and a lattice framework including all the points has been superimposed. Such a lattice is called a stacking lattice. Because all the scattering points lie on this lattice they must all scatter in phase in directions corresponding to the diffracted beams from this lattice. Since the horizontal cell edge of the stacking lattice is half the length of the corresponding cell edge of the lattice of Fig. 4.1*a*, the corresponding horizontal cell edge of the reciprocal stacking lattice must be twice as long as that of the original reciprocal lattice of Fig. 4.1*b*. Thus the sharp spots of Fig. 4.1*b* correspond to diffraction from the stacking lattice.

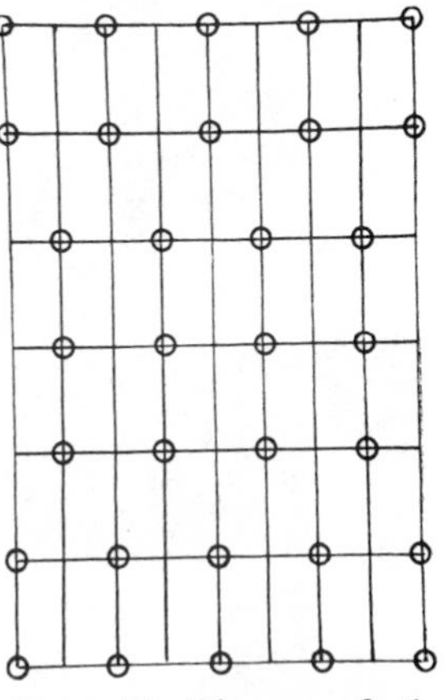

FIG. 4.2 Diagram of the distribution lattice which comprises all the lattice points of Fig. 4.1*a*.

The probability of occurrence of the mistake, consisting of the shift of half the cell side, determines the length and the distribution of intensity in the diffuse spot. Willis (1958) has made a survey of the types of diffraction pattern which can be observed for different values of the quantity α, which is the probability that a fault will occur. A set of four such diffraction patterns is shown in Fig. 4.3. It will be seen that the sharp spots are constant throughout, corresponding to the common stacking lattice. A small probability, e.g. $\alpha = 0{\cdot}1$, corresponds to Fig. 4.1*a*, *b*. When α reaches a value of 0·8 the main part of the structure is face-centred, Fig. 4.4, and the corresponding diffraction pattern is almost face-centred. The sharp spots are the same but the face-centring spots are diffuse. When $\alpha = 0{\cdot}5$ there is an equal probability of either arrangement occurring and the corresponding diffuse spot extends continuously from the top to the bottom of the pattern (Fig. 4.3).

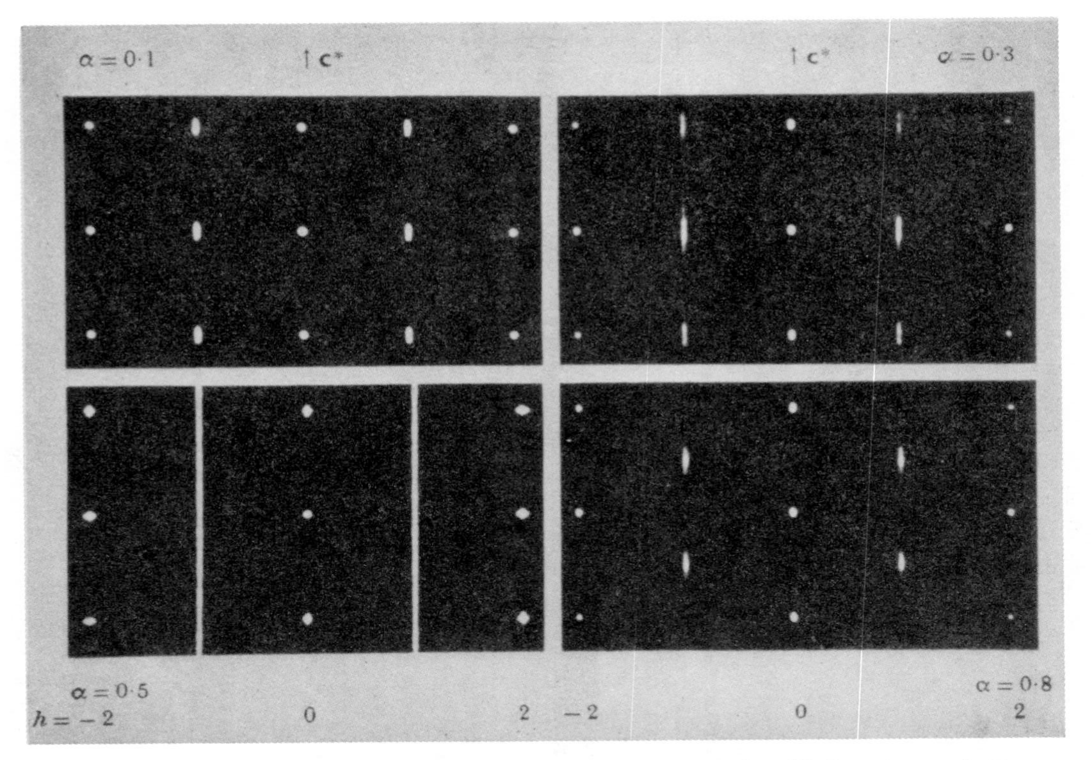

FIG. 4.3 Four optical diffraction patterns corresponding to masks based on Wollastonite-type faulting with probabilities of faulting α = 0·1, 0·3, 0·5, and 0·8 respectively. (After Willis, 1958).

The X-ray photographs of Wollastonite show diffuse spots for indices of the type $h1l$, $h3l$, etc., and they are elongated along **a***. From the above discussion it follows that because the diffuse spots are only diffuse in the direction **a***, the arrangement of atoms must be perfect within any atomic plane parallel to (100). Next we observe that k is *odd* for all diffuse spots and *even* for all sharp spots. This implies that the stacking lattice has a cell dimension of $\mathbf{b}/2$ in the y-axis direction. These two requirements are met by assuming that perfect (100) planes are randomly displaced with respect to one another, the relative displacement of neighbouring (100) planes being always $\pm\mathbf{b}/2$.

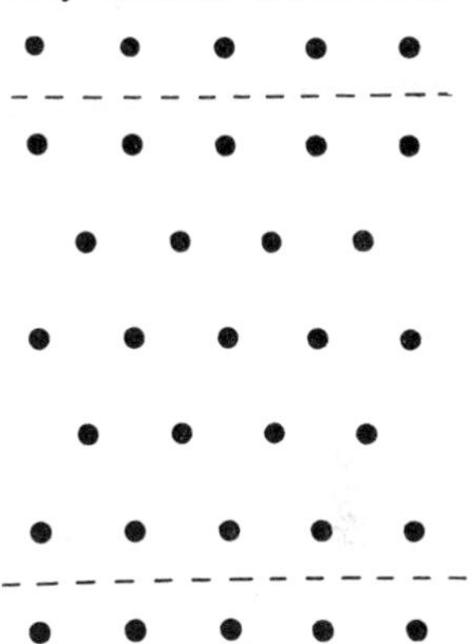

FIG. 4.4 Diagram of Wollastonite-type faulting in which the probability of faulting is very high.

4:2.1.2 *Cobalt*

The stable crystal structure of cobalt at room temperature is hexagonal close-packed. Above 500°C the stable phase is cubic close-packed. Some heat treatments result in an imperfect structure at room temperature. Though the greater part of the crystal is hexagonal, there are thin sheets parallel to the (00·1) plane of the hexagonal structure, in which it is cubic. Mistakes of this kind in the regularity of the structure are possible because both hexagonal and cubic close-packings are obtained by the stacking of identical sheets of atoms arranged on a hexagonal network. Figs. 4.5 and 4.6 show projections of the structures of hexagonal and cubic close-packing on to corresponding planes, viz. (00·1) for h.c.p. and (111) for c.c.p. The thickness of each layer of atoms is the same so that the height of the h.c.p. unit cell is twice the thickness of one layer of atoms while the body diagonal of the unit cell of the c.c.p. is three

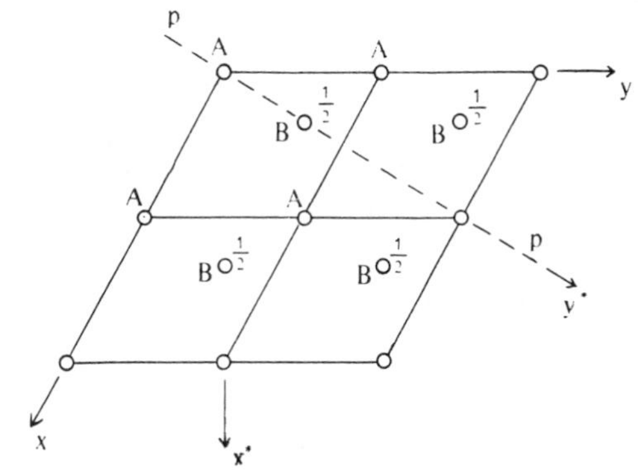

FIG. 4.5 Projection of hexagonal close-packing on (00·1)

times the thickness of one layer. If the successive layers are denoted *A*, *B*, *C* according to their relative positions, the h.c.p. may be described by the sequence *ABAB*... while c.c.p. may be described as *ABCABC*.... The atomic distribution under consideration here is a random combination of these two sequences, for instance *ABACAC*... in which a cubic sequence *BAC* is between hexagonal sequences, *ABA* and *CAC*.

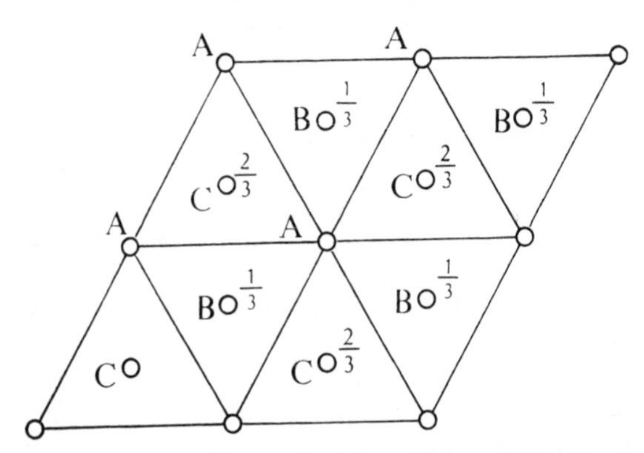

FIG. 4.6 Projection of cubic close-packing on (111) referred to cubic axes or (00·1) referred to hexagonal axes.

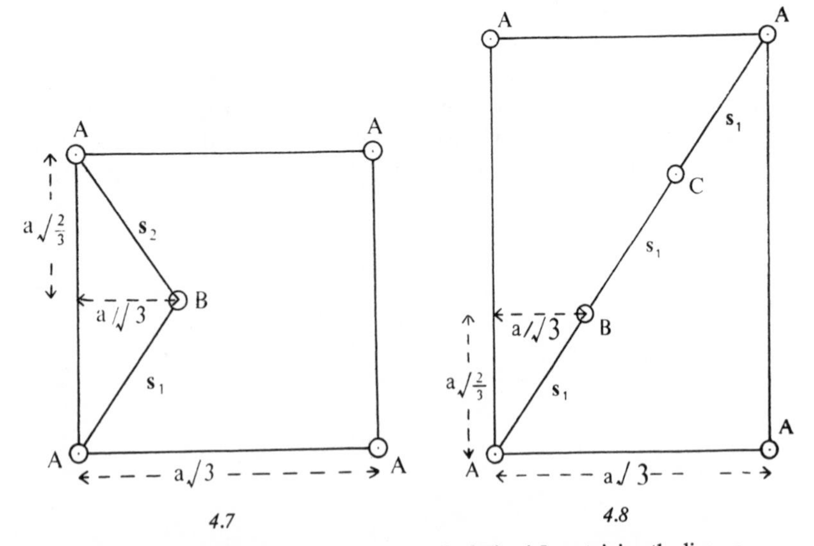

FIG. 4.7 Vertical section of the unit cell of Fig. 4.5 containing the line *pp*.

FIG. 4.8 Vertical section of the unit cell of Fig. 4.6 containing the point row *ABC*.

4:2.1.3 *Stacking vectors*

The relation between the relative positions of the layers may also be expressed in the following way. The sides of the hexagonal unit cell are denoted **a, b, c** respectively. To obtain a *B*-layer from the *A*-layer immediately below it a translation $\mathbf{s}_1$ may be given (Fig. 4.7), where

$$\mathbf{s}_1 = \tfrac{1}{3}\mathbf{a} + \tfrac{2}{3}\mathbf{b} + \tfrac{1}{2}\mathbf{c}.$$

To obtain the next layer above this one a translation $\mathbf{s}_2$ may be given, where

$$\mathbf{s}_2 = -\tfrac{1}{3}\mathbf{a} - \tfrac{2}{3}\mathbf{b} + \tfrac{1}{2}\mathbf{c}.$$

The vectors $\mathbf{s}_1$, $\mathbf{s}_2$ are called *stacking vectors* (Dornberger-Schiff, 1956) and the hexagonal close-packed arrangement of Fig. 4.5 corresponds to a succession $\mathbf{s}_1, \mathbf{s}_2, \mathbf{s}_1, \mathbf{s}_2$ (Fig. 4.7). The cubic close-packing of Fig. 4.6 corresponds to a succession $\mathbf{s}_1, \mathbf{s}_1, \mathbf{s}_1, \ldots$ (Fig. 4.8). The random combination of hexagonal and cubic close-packing *ABACAC* may be described by the sequence of stacking vectors e.g. $\mathbf{s}_1, \mathbf{s}_2, \mathbf{s}_2, \mathbf{s}_1, \mathbf{s}_2, \ldots$.

In Fig. 4.9 this sequence is drawn through the points. Such points are called *representative points* because such a single row of points when repeated

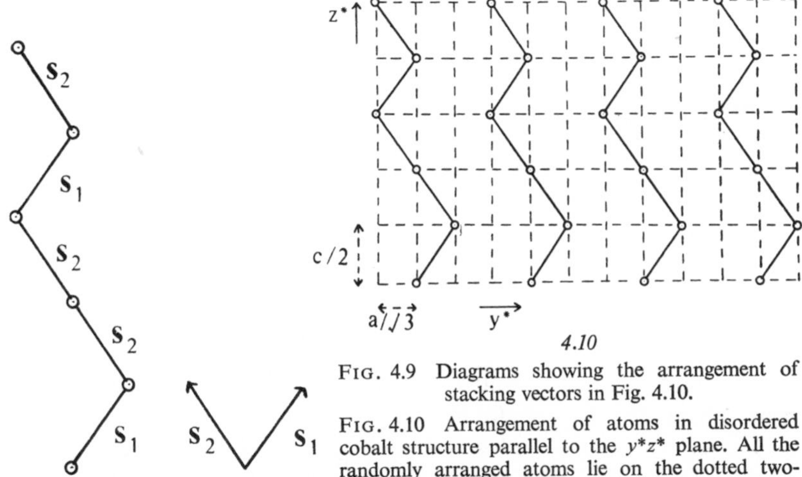

FIG. 4.9 Diagrams showing the arrangement of stacking vectors in Fig. 4.10.

FIG. 4.10 Arrangement of atoms in disordered cobalt structure parallel to the y^*z^* plane. All the randomly arranged atoms lie on the dotted two-dimensional distribution lattice.

in a hexagonal network generates the whole distribution of atoms. All the stacking vectors are parallel to the y^*z^* plane of the reciprocal lattice. The arrangement of atoms parallel to this plane is shown in Fig. 4.10. The whole structure is composed of planes of atoms such as that shown in Fig. 4.10 placed side by side so as to form the hexagonal network in the (00·1) plane. The atomic layers of Fig. 4.10 do not form a regular two-dimensional lattice since the stacking vectors $\mathbf{s}_1$, $\mathbf{s}_2$ follow one another in a random sequence. However, all the atoms lie on the *distribution lattice* shown by the dotted lines in Fig. 4.10. This is rectangular and has cell dimensions one-third of the original hexagonal cell in the y^*-direction and one half in the z^*-direction. We shall now consider the diffraction effects which the hexagonal network of atoms parallel to (00·1) and the not-truly-periodic network parallel to $(2\bar{1}\cdot 0)$ give separately and in combination.

4:2.1.4 *Diffraction effects*

The diffraction effect due to the whole crystal may be analyzed into the effects of identical layers which are arranged parallel to one another. The individual layers of hexagonal symmetry which are parallel to (00·1) in the hexagonal phase, and to (111) in the cubic phase, have a reciprocal equivalent

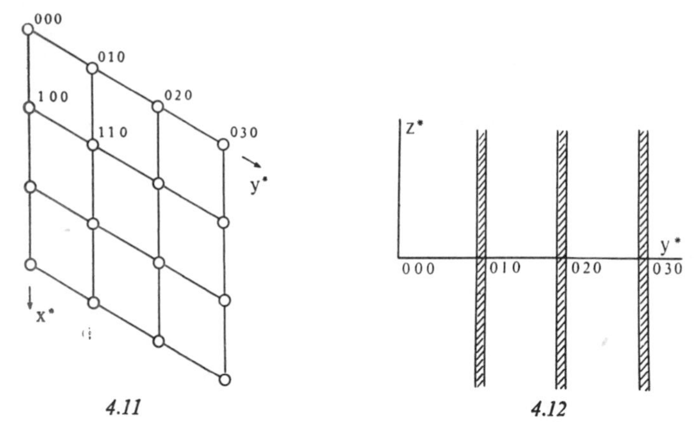

FIG. 4.11 The circles represent diffuse rods parallel to z^* passing through the corners of the reciprocal unit cell.

FIG. 4.12 The shaded areas represent the same diffuse rods as those denoted by small circles in Fig. 4.11.

consisting of rods parallel to the z^* hexagonal axis or to the [111] axis of the cubic phase. This feature also characterizes all the disordered structures in which the layers are shifted irregularly into the positions denoted *A*, *B*, *C* in Figs. 4.5 and 4.6. Two sections of reciprocal space showing these rods are given in Figs. 4.11 and 4.12. The rods pass through the corners of the base of the hexagonal reciprocal unit cell and diminish slowly in intensity with distance from the equatorial plane. The layers of atoms parallel to the plane $(2\bar{1}\cdot 0)$ give a diffraction pattern which, although not the same as that given by the corresponding two-dimensional distribution lattice of Fig. 4.10, has some features in common with it. The diffraction due to the distribution lattice consists of continuous rods perpendicular to the plane $(2\bar{1}\cdot 0)$ and passing through points in this plane which are indicated by open circles in Fig. 4.13. From Fig. 4.5 it is clear that the projection of the line *AB* on to the (00·1) plane is 2/3 of the perpendicular distance between the opposite sides of the unit cell. Hence the first point along y^* is a distance of $3a^*/2$ from the origin, as shown in Fig. 4.14. Similarly, in Fig. 4.7 the height of the unit cell of the distribution lattice may be seen to be c/2. Hence the first reciprocal distribution point along z^* is $2c^*$ from the origin as in Fig. 4.13. In Fig. 4.14 is shown a plan of the diffuse rods associated with the distribution lattice. It will be seen that not all points of the distribution lattice are filled and therefore there will be some diffracted intensity along lines between those shaded in Fig. 4.14. This part of the diffracted intensity is associated with the diffuse spots produced by the actual crystal.

The whole structure may be regarded as being formed by the repetition

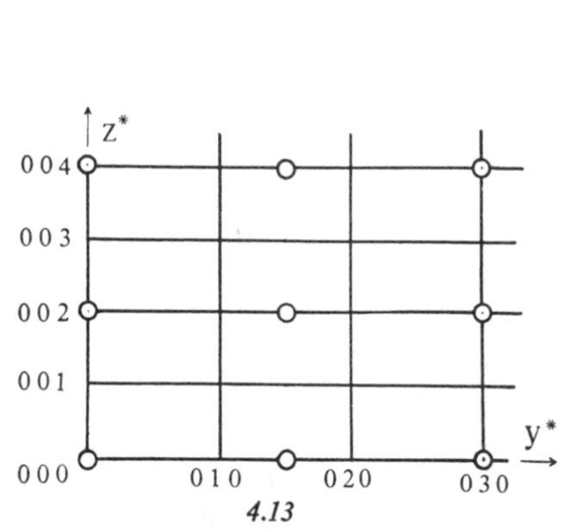

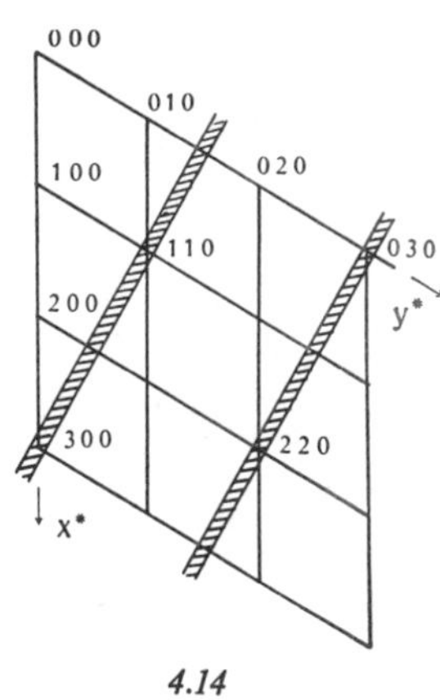

FIG. 4.13 The circles represent the sections by the y^*z^* plane of the same diffuse rods as those indicated by the shaded areas of Fig. 4.14.

FIG. 4.14 The shaded areas represent diffuse rods perpendicular to the plane containing the stacking vectors $\mathbf{s}_1$, $\mathbf{s}_2$.

at each net point of the hexagonal net *AAAA* . . . (Fig. 4.5) of the sequence of stacking vectors $\mathbf{s}_1$, $\mathbf{s}_2$, $\mathbf{s}_2$, $\mathbf{s}_1$, $\mathbf{s}_2$, ... (It is understood that there is an atom placed at the end of each stacking vector.) From this way of regarding the structure we can see that *only along the vertical rods* in reciprocal space shown in Figs. 4.11 and 4.12 can there be any scattering at all, i.e. both sharp and diffuse reflections must occur in these rods. An alternative way of regarding the structure is as a repetition of the distribution lattice of Fig. 4.10 at all the points *AA* . . . (Fig. 4.5) along the *y*-axis. From our previous discussion it is clear that all atoms will scatter in phase along the shaded reciprocal rods of Fig. 4.14. Thus at the intersection of the rods of Figs. 4.11 and 4.14 all atoms will scatter in phase and sharp reflection will consequently be observed.

As may be seen by inspecting the Figs. 4.11 and 4.14, this occurs at reciprocal points for which

$$h-k = 3n$$

and

$$l = 2m,$$

where n and m are integers. When these conditions are not fulfilled the reflections may be diffuse and drawn out along the vertical direction perpendicular to the hexagonal nets.

The position in which relps would occur if the stacking vectors had a cubic sequence is shown, for the y^*z^* plane, in Fig. 4.15. The rectangular framework is that of the hexagonal crystal; the large (open) circles and the small (filled) circles correspond to the relps of the corresponding cubic sequences $s_1s_1s_1 \ldots$ and $s_2s_2s_2 \ldots$ respectively. The large and small circles coincide at the relps corresponding to sharp spots and are separated from the hexagonal relps which have diffuse spots.

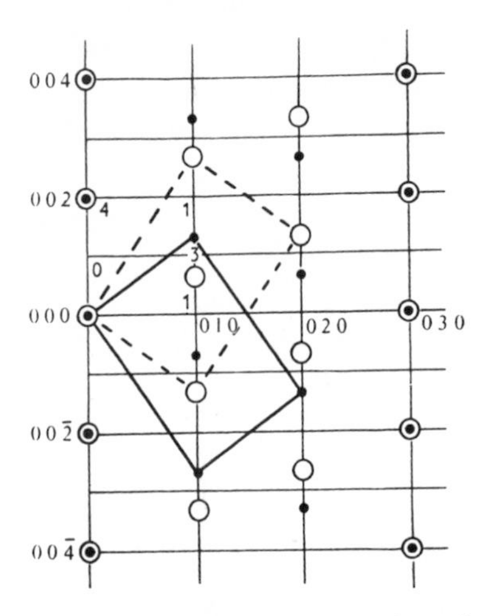

FIG. 4.15 A section of the reciprocal lattice of cubic cobalt with hexagonal indices. The rectangle bounded by dotted lines gives a (110) section of the cubic cell in one of the two possible orientations and the rectangle bounded by full lines gives the alternative orientation.

As in the thermal diffuse spots so here also the intensity of the diffuse spot is proportional to the intensity of the corresponding normal Bragg reflection. The numbers 0, 1, 3, 4 in Fig. 4.15 give the calculated F^2_{hkl}-values for the corresponding hexagonal lattice points. The intensities of the diffuse spots centred on 010 and 011 are in the ratio 1:3, which is the same as the corresponding ratio of their F^2_{hkl}-values. The diffuse spots for l_{even}, are more elongated than those for l_{odd}, and from the breadth of these diffuse spots Wilson (1949*a*) has calculated the probability of an interruption in the correct hexagonal sequence. Willis (1958) made an optical diffractogram, Fig. 4.16, based on the model discussed here and it shows the same features as are observed by diffuse X-rays.

4:3 **Diffuse scattering from diamond crystals**

4:3.1 *Experimental results*

The diffuse reflections given by Type I diamonds have been studied by photographic and ionization methods. A Laue photograph such as Fig. 4.17

shows the three sharp spots *a*, *b*, *c*, corresponding to the intersection with the reflecting sphere of the spikes parallel to [100], [010], and [001], starting from the 111 relp. This relp gives the normal Laue spot *L*. The directions of these spikes have been studied by taking a succession of Laue photographs in which the crystal has been rotated about a given axis, usually [110], through an angle of about 1° between each exposure.

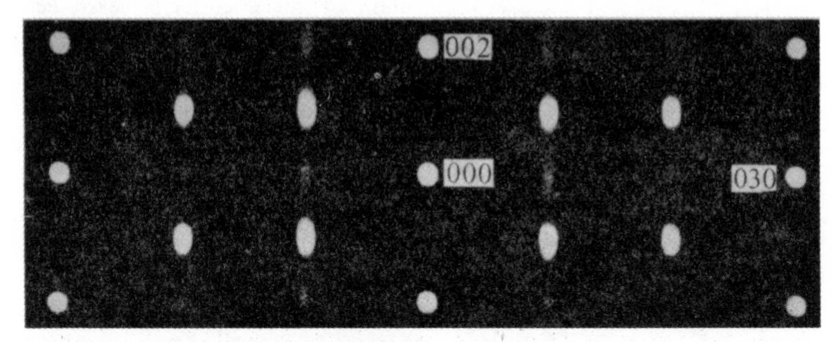

FIG. 4.16 Optical diffraction pattern given by a mask representing growth-faulted hexagonal cobalt. (After Willis, 1958).

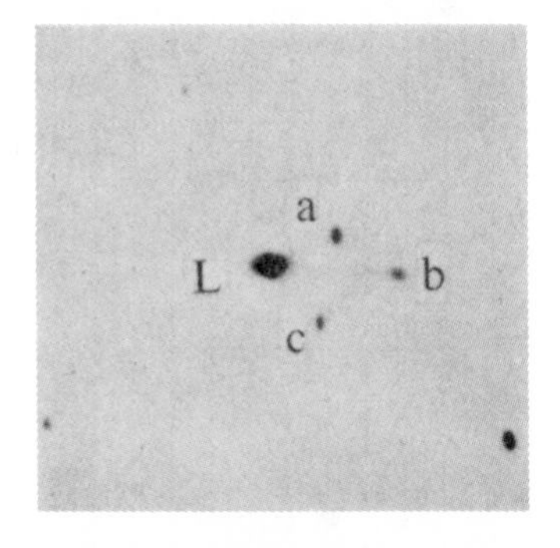

FIG. 4.17 Laue photograph of Type I diamond, showing the spots (*a*) (*b*) (*c*) due to the spikes associated with the relp giving the Laue reflection *L* of indices 111.

The intensity variation along the spikes has been studied by calibrating the intensity of the spike spot against the Bragg reflection of the characteristic radiation. For this purpose Hoerni and Wooster (1955) used a combination of Laue and Weissenberg photographs. The intensity was found to decrease with distance from the relp according to an exponential law, the value of the exponent being 2.2. If we consider only spikes which are parallel to [100], the following statement summarizes the relation between their intensities and the indices *h*, *k*, *l* of the relp from which they arise. The intensities of [100] spikes depend only on the values of *h* and are independent of the values of *k* and *l*. For the *h*-values 1, 2, 3, and 4 the relative intensities of the spikes are 100, 75, 5, and 30 respectively. (These observations were made on a relatively small number of crystals and cannot claim to be widely representative.) Exactly similar statements can be made about the spikes parallel to [010] and

[001]. Fig. 4.18 shows a summary of the spike intensities for the spikes parallel to [010]. The remarkable feature of this intensity distribution is the falling to zero, or near zero, of the spike intensities associated with the index 3. Thus, for spikes parallel to [100] the intensity is nearly zero whenever $h = 3$, and for spikes parallel to [010] and [001] the intensity is nearly zero when k or l respectively is equal to three.

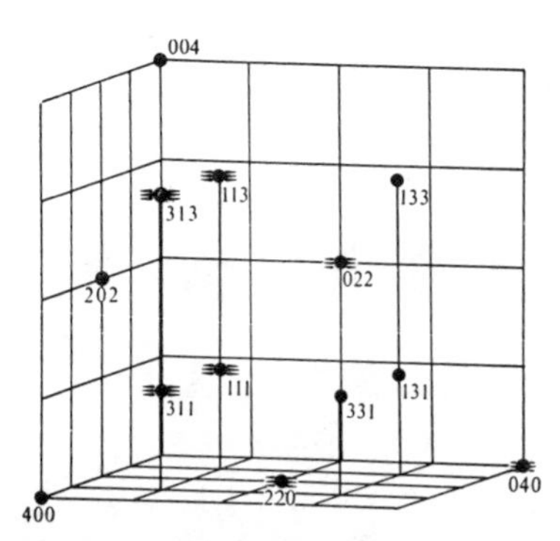

FIG. 4.18 The spikes parallel to [010] associated with the reciprocal lattice points of diamond.

4:3.2 *A model of an occasionally twinned structure*

A formal explanation of these observations is easily given (Wooster, 1956), though one that is generally acceptable has not so far been put forward. Let us first consider the explanation for one set of spikes, say, those parallel to [100]. It is clear that any stratification of the crystal structure parallel to (100) will result in spikes normal to this plane. Further, if the misplacements which

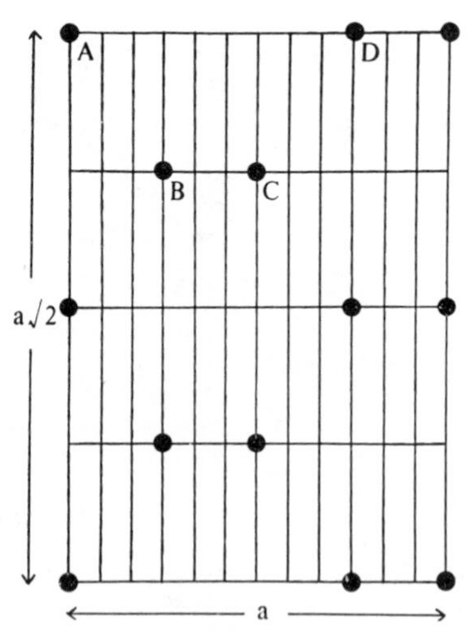

FIG. 4.19 A projection of one unit cell of the diamond structure on a $(01\bar{1})$ plane. The rectangular lattice is a distribution lattice for the reflection-twin on (100).

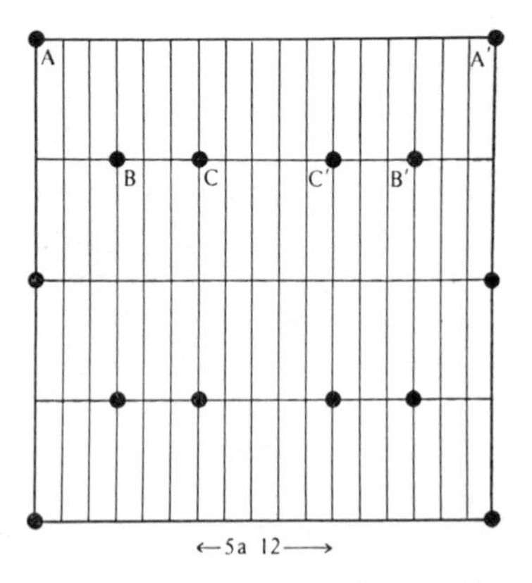

FIG. 4.20 Diagram showing the same projection of the diamond structure as in Fig. 4.19 but with the right-hand half of the pattern twinned with respect to the left-hand half. Atoms in both halves of the pattern are on the same distribution lattice.

arise between one perfect stratum and the next occur at random the decrease of spike intensity according to an approximate inverse square law will be expected (Wilson 1949*a*). The zero intensity for $h = 3$ implies that the distribution lattice is based on a cell of side $a/3$. Bearing in mind that along the [100] axis equivalent planes are separated by distances of $a/4$, the distribution lattice which contains all the atoms of the structure must be based on a cell side of $a/12$ in the [100] direction.

In order to make an optical analogue which gives diffraction effects associated with planes such as (111), (311), etc., a plan of the diamond structure on the plane $(01\bar{1})$ is made, Fig. 4.19. There are four atomic positions in each unit cell represented by the points *A*, *B*, *C*, *D*, respectively. Distribution lattice lines separated by a distance of $a/3$ must pass through each of these points and this divides the cell into twelve parts along the [100] axis, as shown by the lines of Fig. 4.19. There are several ways in which atoms can be arranged on such a distribution lattice and only one of them is put forward here. This arrangement, which is a reflection-twin, is shown in Fig. 4.20. With the same distribution lattice as that shown in Fig. 4.19 the atoms *A*, *B*, *C* are in the original positions and those marked *A'*, *B'*, *C'* are in twin positions related to the original positions by a reflection in a (100) plane midway between *C* and *C'*. Based on this relative arrangement of primed and unprimed lattice

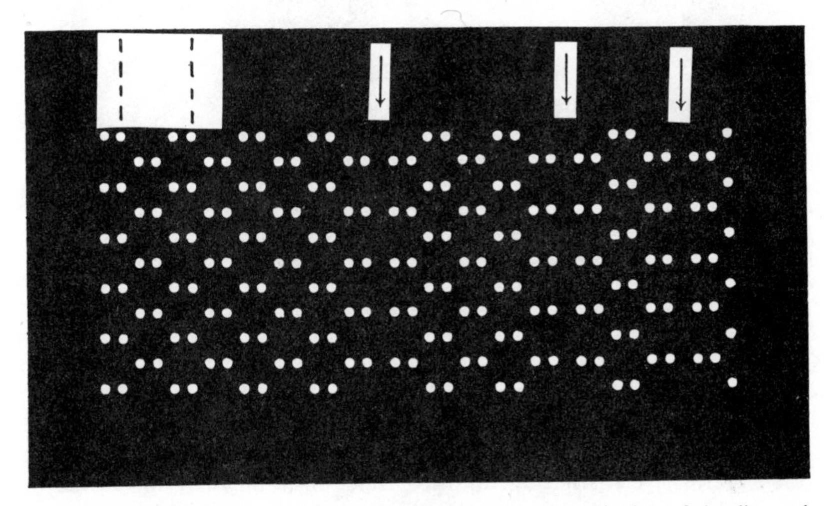

FIG. 4.21 Diagram of optical mask representing the same projection of the diamond structure as Fig. 4.19. The three arrows show planes of reflection twinning.

points a mask is made according to Fig. 4.21, from which it will be seen that there are three such reflection planes marked with arrows. The repeating unit corresponding to Fig. 4.19 is shown by dotted lines. It will be seen that normal to the axis [100] the arrangement is that of perfect strata and only along the axis [100] are there three reflection-twin planes which always bring the atoms on to the same distribution lattice. The diffraction pattern obtained with the mask of Fig. 4.21 is shown in Fig. 4.22. It will be seen that the reflections 111,

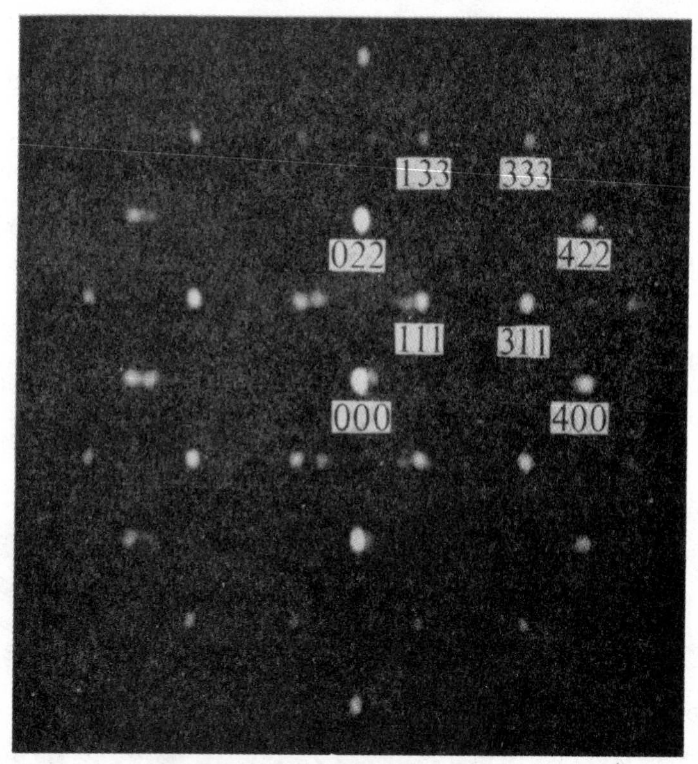

FIG. 4.22 Diffraction pattern of diamond structure projected on the plane $(01\bar{1})$, obtained with the mask of Fig. 4.21.

133, 400, 422 have diffuse spikes but that 311, and 333 are sharp reflections. (The diffuse spikes do not extend symmetrically on either side of the relp but that is due to the limitation of the model. Compression of the matrix containing the twinned layers is neglected.) This result, which may be summarized by saying that spikes are observed when $h = 1$ or 4 but not when $h = 3$, corresponds with the experimental result that the intensities of the spikes are in the ratios 100, 5, and 30 for $h = 1$, 3, and 4 respectively. The arrangement of the atoms on the distribution lattice here chosen necessarily results in sharp spots for $h = 3$. The atom C in Fig. 4.20 is bound to only three neighbours, namely C' on the right and two B atoms on the left. These bond distances are of practically the same length, being $0{\cdot}42a$ $(a.5/12)$ and $0{\cdot}43a$ $(a.\sqrt{3}/4)$ respectively. Such a triangular coordination round C and C' does not accord with the usual ideas concerning tetrahedral coordination of carbon in diamond. However, it seems possible that nitrogen atoms, which often occur as an impurity in diamond, may occur in such positions as CC' (Elliott, 1960). Platelets of nitrogen atoms of thickness corresponding to CC' afford the most probable explanation of the spikes so far advanced.

The discussion up to this point has been concerned with spikes parallel to the axis [100] only. In fact the spikes occur parallel to all the three axes [100], [010], and [001], subject only to the limitations already given. Two explanations may be advanced to account for this. In Fig. 4.23*a* is shown a domain structure in which neighbouring portions of the crystal have the twin planes of Fig. 4.20 arranged parallel to only one of the three planes (100), (010), and (001). If these domains are on a scale of, say, 0·01 mm they would be large enough to account for the form of the spikes and numerous enough to account for the equal intensities of the corresponding spikes parallel to the three axes [100], [010], and [001]. An alternative structure is shown in Fig. 4.23*b* in which the

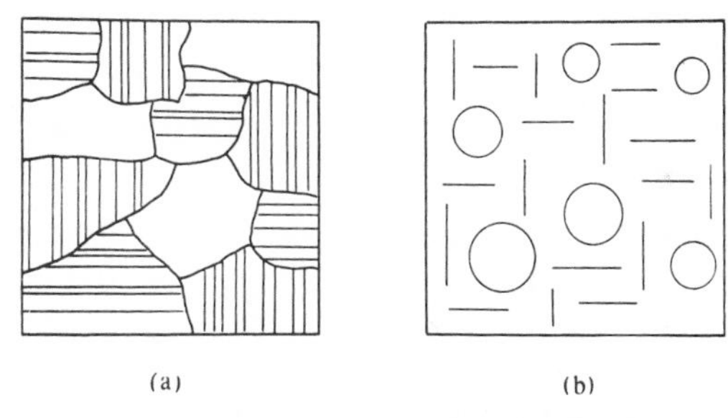

FIG. 4.23 Diagrams showing (*a*) a domain structure with spike directions along all three cubic axes in different domains; (*b*) a random arrangement of (100) planes (denoted by lines and circles) across which the twinning shown in Fig. 4.20 occurs.

twin planes represented by lines and circles do not extend right across the crystal but are nevertheless large enough to give spikes of well defined cross section. Metallurgical photographs often show precipitates which have this type of arrangement. Up to the present, experimental work has not indicated which of these alternatives corresponds with actual atomic arrangements in diamond.

4:3.3 *The density of twin planes*

The last problem concerns the frequency with which the structural faults occur. An estimate of this can be made from the measurement of the absolute intensity of the spike, i.e. the intensity expressed as a fraction of the intensity of the incident beam. This measurement was carried out on three crystals (Hoerni and Wooster, 1955) by comparing the intensity of the spikes on $\bar{3}11$ parallel to [010] and [001] with two points of the thermal diffuse spot (near to $\bar{3}11$ Laue spot), corresponding to the rekhas parallel to [110] and [11$\bar{1}$]. From the known elastic constants the thermal diffuse intensities, expressed as a fraction of the intensity of the incident beam, could be calculated. Thus the intensity at any desired point along the spike could be determined first in relation to the thermal diffuse intensity and then in relation to the intensity of the incident beam. The experimental result can be expressed in the following way. For the crystal showing the strongest spike we suppose the incident beam to be completely intercepted by a plane face ($\bar{3}11$). Then the ratio of the flux at the spike spot, at a distance $1/10a$ from the relp, to the incident flux is

$$I/I_0 = 1{\cdot}15 \times 10^{-6}.$$

Wilson (1949*a*) has given a theory of the diffuse scattering from a structure with defects in parallel layers. If α is the probability of a defect occurring and ζ is the distance from the relp expressed as a fraction of the side of the reciprocal cell, then, neglecting squares of α, we have

$$I/NJ_0 = \alpha/2\pi^2\zeta^2,$$

where I is the diffuse flux at the point defined by ζ and NJ_0 is the intensity of the Bragg reflection. Combining the results of I/I_0 given above with those independently measured for NJ_0/I_0, the value of α was found to be $2{\cdot}6 \times 10^{-4}$. This work was repeated by Caticha-Ellis and Cochran (1957) using a counter diffractometer. An octahedral diamond exhibiting strong spikes was set to give a spike lying in the equatorial plane associated with a 111 relp. The variation of the intensity of the spike along its length and also its absolute intensity were determined. The Bragg integrated reflection was directly measured and corrected for extinction by comparison with reflections from LiH. The value of α so determined was $4{\cdot}6 \times 10^{-4}$ (± 20 per cent.) or, approximately, one fault every 2,000 atomic planes.

Hoerni and Wooster had shown that the spike intensity varied along its length exponentially and that the exponent was $2{\cdot}2 \pm 0{\cdot}1$. Caticha-Ellis and Cochran pointed out that this did not conflict with the theory due to Wilson, which requires the second power for the exponent of ζ, because the factor F_{hkl}^2 also varies near to the relp. As was shown in equation (1.9), the diffuse intensity due to thermal waves depends on F_{hkl}^2. The same is true here but it must be remembered that the relevant value of F_{hkl}^2 is that which applies at the point on the spike under consideration. When this is taken into account the experimentally determined exponent does not differ significantly from 2.

4:4 Diffuse reflections given by $AuCu_3$

The alloy $AuCu_3$ may be crystallized in a regular structure, one unit cell of which is shown in Fig. 4.24. If the difference between copper and gold

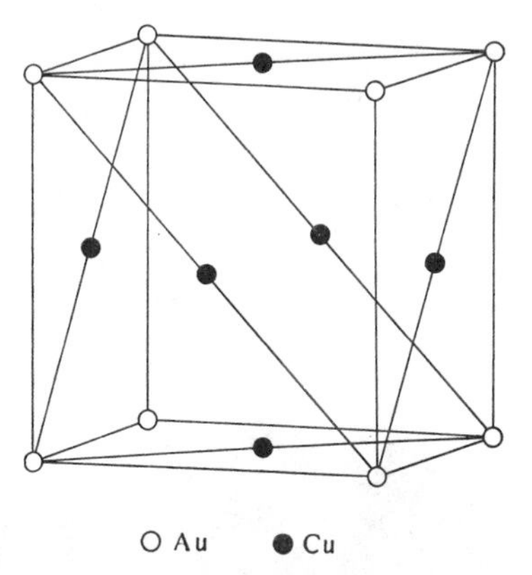

FIG. 4.24 Diagram showing the atomic arrangement in the alloy $AuCu_3$ in the fully ordered state.

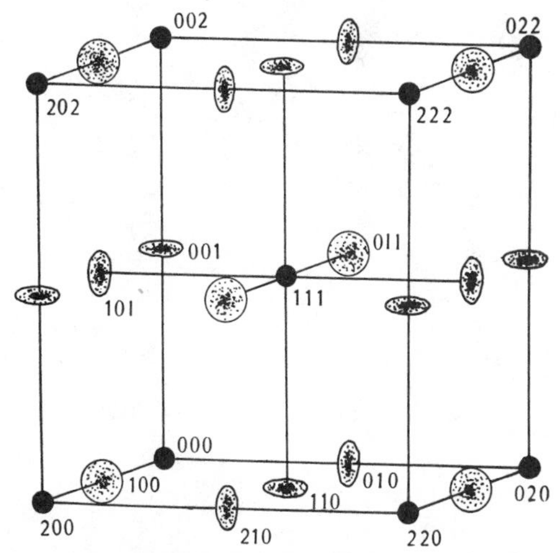

FIG. 4.25 Diagram in reciprocal space showing sharp relps having indices all even or all odd, and diffuse disk-like relps having mixed indices.

atoms is ignored the structure is face-centred cubic. If account is taken of this difference the structure is based on a primitive cubic lattice and gives X-ray reflections of any indices. Such a perfect structure can only be obtained by

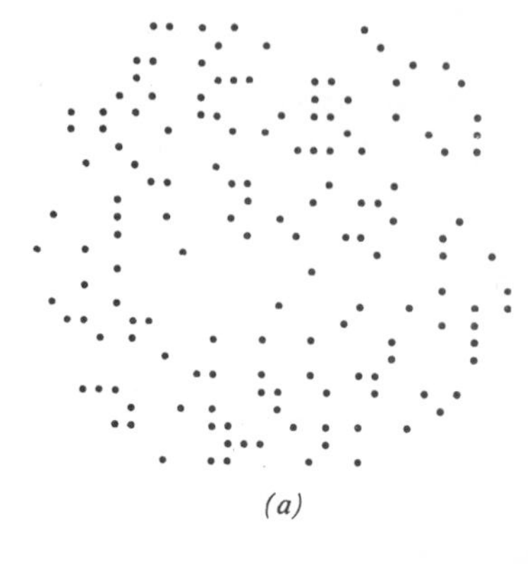

(a)

FIG. 4.26*a* Diagram of optical mask corresponding to a projection on the (100) plane of the random arrangement of (Au—Cu) 'atoms' on the face-centred cubic lattice.

FIG. 4.26*b* Diffraction pattern given by the mask of Fig. 4.26*a*.

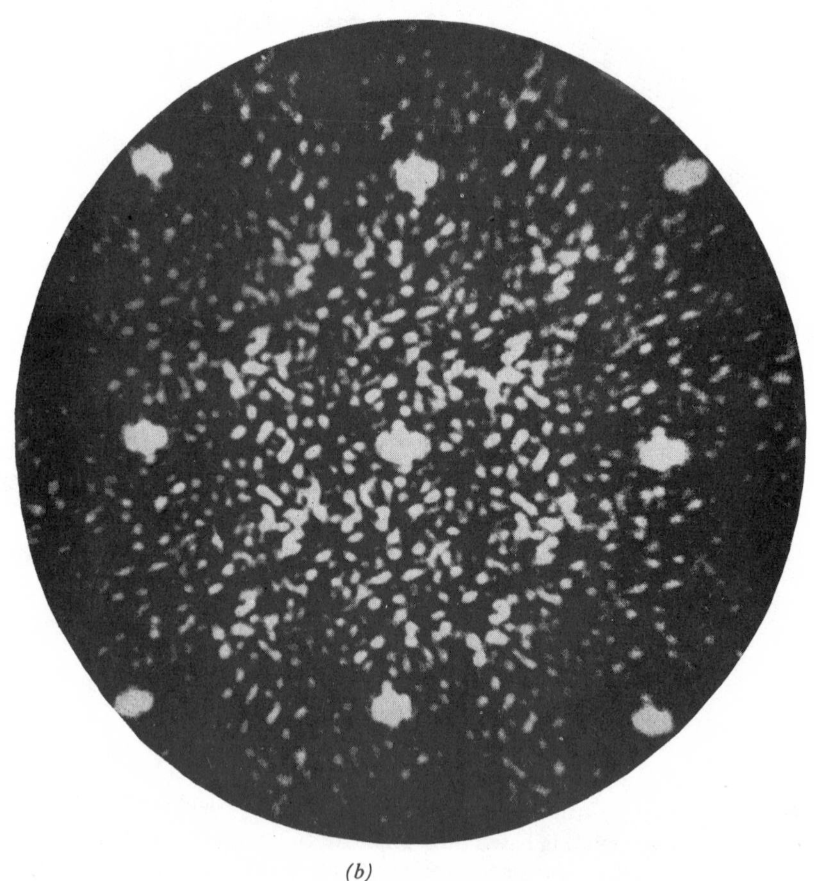

(b)

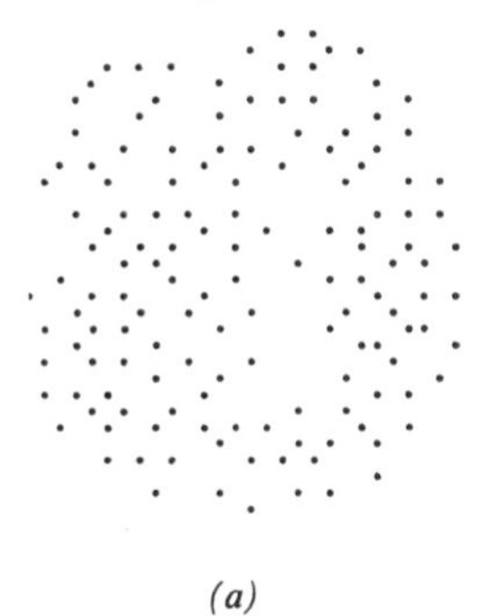

(a)

FIG. 4.27*a* The same as Fig. 4.26*a* but with points moved further apart though still on the same partial face-centred cubic lattice.

FIG. 4.27*b* Diffraction pattern given by the mask of Fig. 4.27*a*.

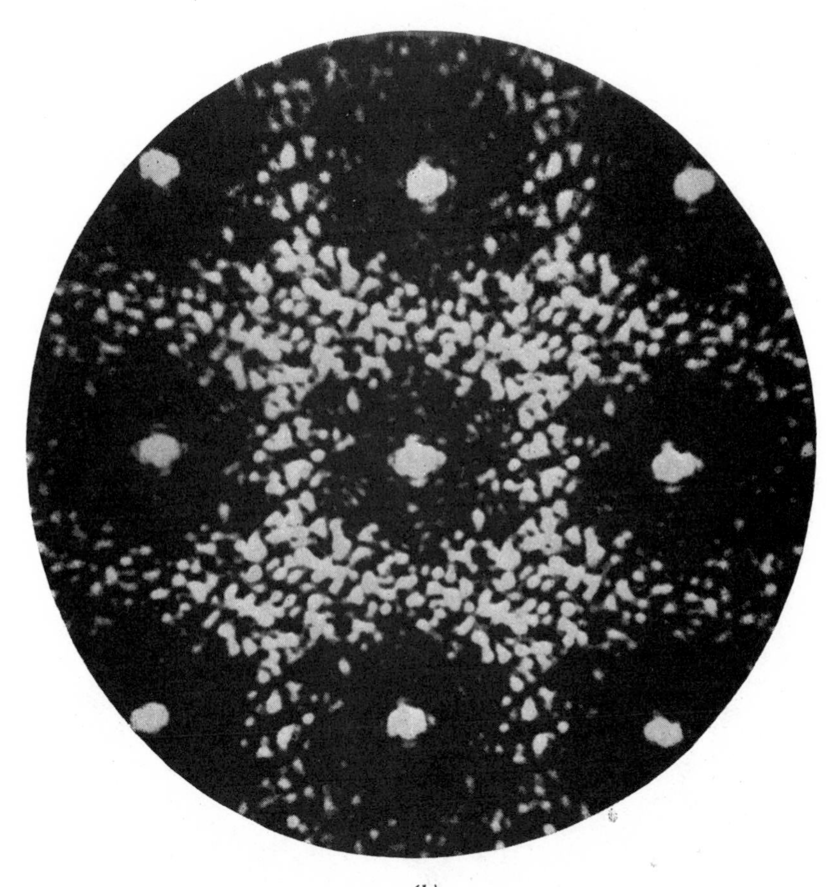

(b)

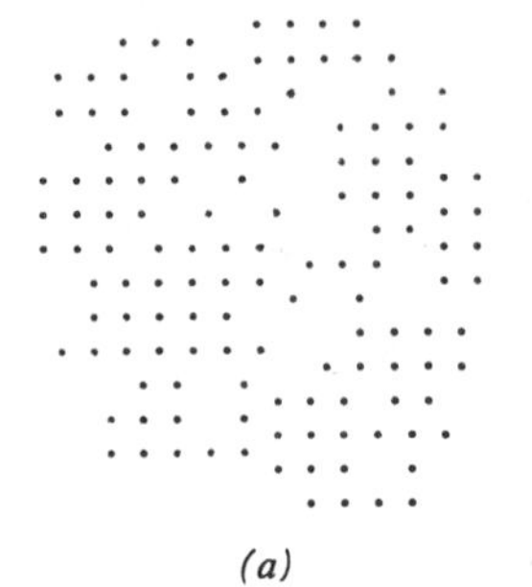

(a)

FIG. 4.28*a* The same as Fig. 4.27*a* but with still more order introduced into the arrangement of points.

FIG. 4.28*b* Diffraction pattern given by the mask of Fig. 4.28*a*. Compare this with Fig. 4.25.

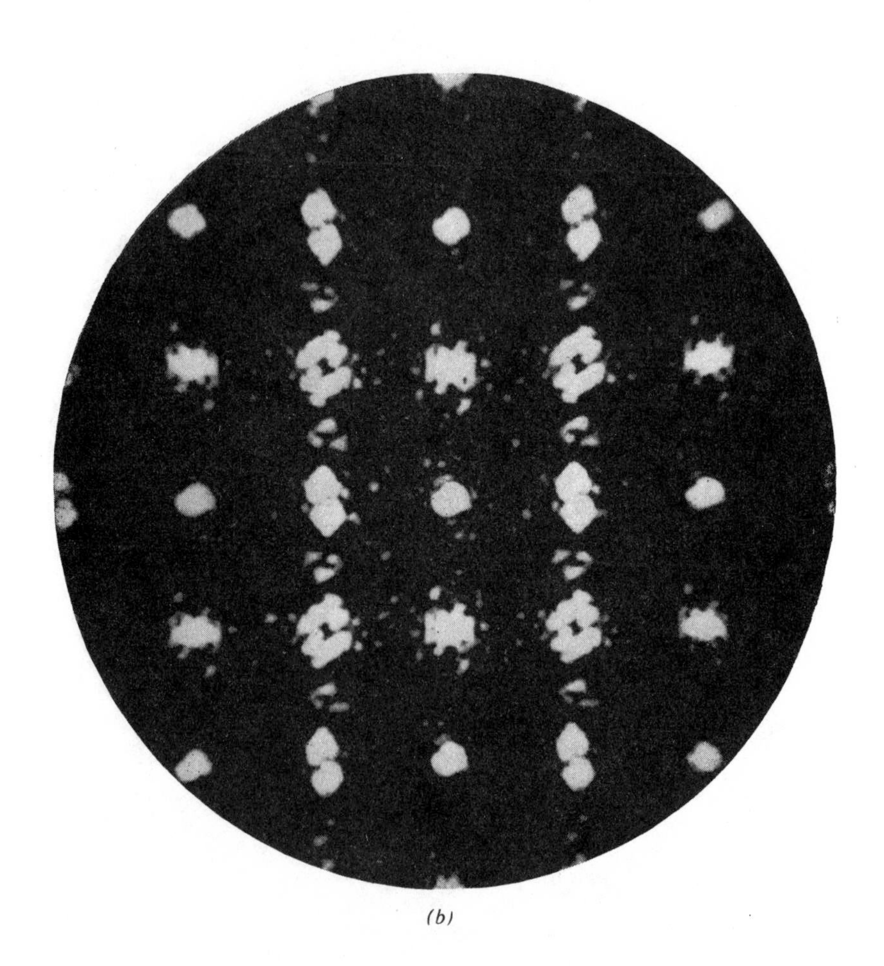

(b)

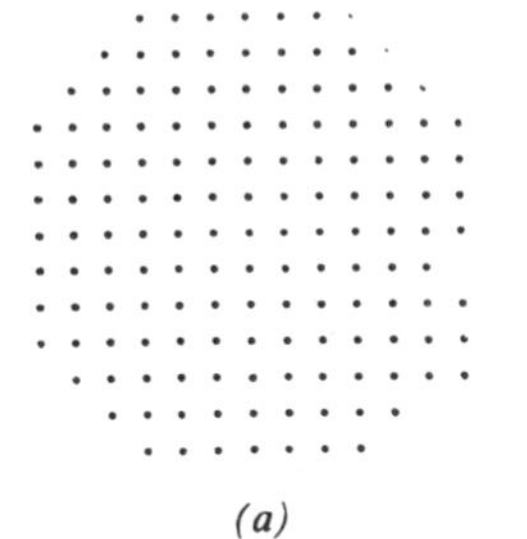
(a)

FIG. 4.29*a* Fully ordered arrangement corresponding to that shown in Fig. 4.24.

FIG. 4.29*b* Diffraction pattern given by the mask of Fig. 4.29*a*.

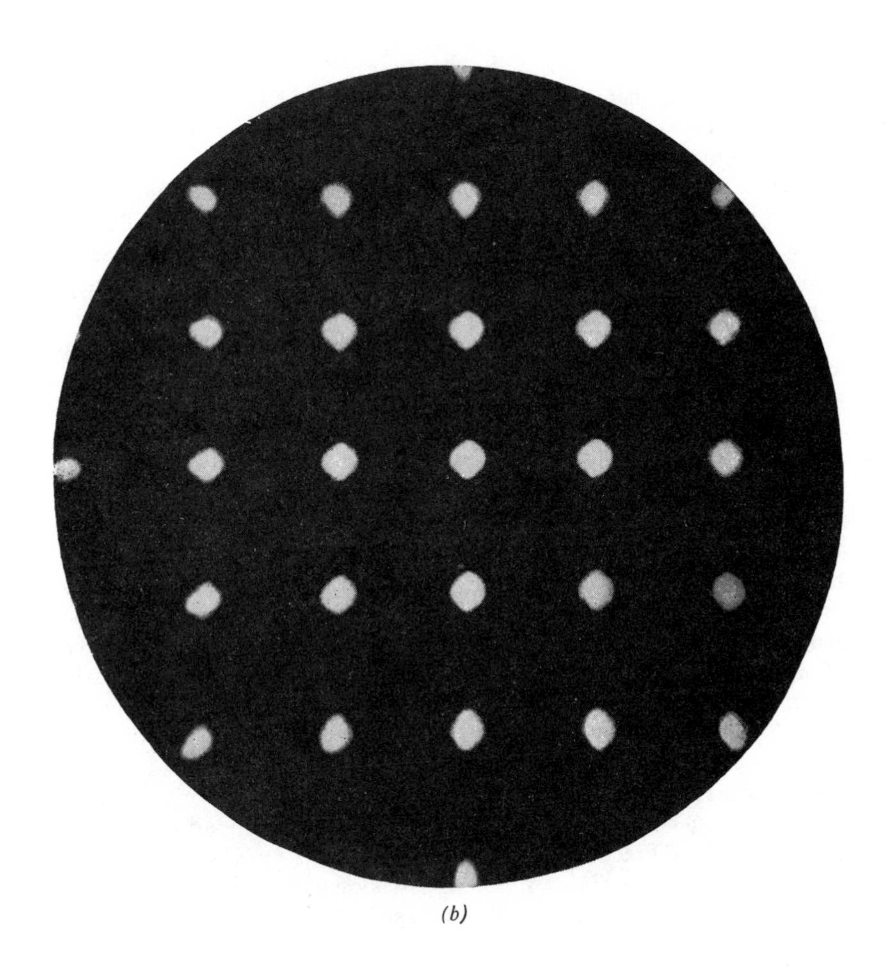
(b)

prolonged annealing and the structure is usually more or less disordered. The disordered structure gives rise to diffuse reflections (Edmunds, Hinde, and Lipson, 1947), which may be described by Fig. 4.25. In this figure reciprocal points such as 111, 200, 220, etc., which correspond to reflections from the face-centred lattice, are sharp. This is to be expected since the face-centred lattice is the distribution lattice on which all the atoms are located. All other relps, e.g. 100, 110, 210, etc., are diffuse. If a line is drawn through any sharp relp parallel to one of the edges of the unit cell it passes through a neighbouring sharp relp and midway between these sharp relps is a diffuse disk, the plane of which is perpendicular to this line.

An explanation of these experimental results can be given in the following way. Both the disordered and the ordered structures may be regarded as formed by the superposition of two structures. On to the face-centred cubic lattice of copper is superposed a face-centred lattice of the same cell size composed of atoms having a scattering power of $f_{Au} - f_{Cu}$. This second lattice has three times as many vacancies as filled points. In the disordered structure the (Au – Cu) 'atoms' are randomly distributed on the face-centred lattice, whereas in the ordered structure these 'atoms' occur only at the corners of the unit cells. The distribution lattice contains all the points of both superposed lattices and its reciprocal points correspond to the sharp reflections. Because the distribution lattice is face-centred cubic the sharp reflections are those for which the indices are all odd or all even, as shown in Fig. 4.25. All other relps are made diffuse by the randomness of the arrangement of the (Au – Cu) atoms on their face-centred lattice. To study these diffuse reflections we need only consider the partial lattice of the (Au – Cu) atoms.

An illustration of the formation of these diffuse relps can be obtained from an optical analogue. The masks and diffraction patterns of Fig. 4.26 to 4.29 have been prepared by C. A. Taylor (see also Taylor, Hinde, and Lipson, 1951) and represent the projection on a cube face of the (Au-Cu) 'atoms' only. This is justifiable because the face-centred Cu lattice would only contribute to the sharp spots. The masks of Figs. 4.26*a*, 4.27*a*, and 4.28*a* represent arrangements with increasing degrees of order and Fig. 4.29*a* corresponds to the fully ordered structure, the unit cell of which is taken to be of length *a*. Fig. 4.26*a* shows a mask with a random distribution of points on the face-centred lattice of side *a*. The diffraction pattern, Fig. 4.26*b*, has sharp spots on a square lattice and a more or less uniform distribution of scattering elsewhere. In Fig. 4.27*a* the mask represents the distribution in which some of the (Au – Cu) atoms have moved further from one another than they are in Fig. 4.26*a*. (The distance apart of atoms in contact is $a/\sqrt{2}$ and in the mask this corresponds to dots $a/2$ apart in horizontal and vertical directions and $a/\sqrt{2}$ apart in diagonal directions.) The diffraction pattern, Fig. 4.27*b*, shows the same sharp spots as Fig. 4.26*b* but the diffuse scattering is concentrated along vertical and horizontal lines midway between the sharp spots.

This result can be understood if Fig. 4.27*a* is compared with Fig. 4.3 for Wollastonite. It is possible to isolate from Fig. 4.27*a* a set of points which are all spaced at a distance of a or a multiple of a apart along horizontal lines which are themselves a apart. Further, the displacement of successive horizontal rows relative to one another occurs in a random manner and the amount of the displacement is $a/2$. Corresponding to such a distribution we have vertical diffuse rods midway between the sharp spots. It is also possible to choose similar rows of points lying along vertical lines which are separated by a distance a from one another. This gives rise to the horizontal diffuse rods in Fig. 4.27*b*. Thus we can account for the main features of Fig. 4.27*b* by regarding Fig. 4.27*a* as a two-dimensional disordered structure which corresponds in one dimension to Wollastonite.

The separation of the (Au – Cu) atoms from one another is carried further in Fig. 4.28*a*. No two points are closer than a in this figure. The diffraction pattern of this mask is shown in Fig. 4.28*b* and it will be seen that the diffuse regions have contracted. Where the former continuous horizontal and vertical rods crossed one another we have a more or less circular diffuse spot and where the rod, either horizontal or vertical, ran between two sharp spots we have a more or less elliptical diffuse spot, the long axis of the ellipse being perpendicular to the line joining the neighbouring sharp spots. By comparing Fig. 4.28*b* with Fig. 4.25 it will be seen that Fig. 4.28*b* is similar to the plane through the origin 000 parallel to a face (100) in Fig. 4.25.

Finally, in Fig. 4.29*a* is shown the mask corresponding to the ordered alloy in which all gold atoms are at the corners only of the unit cells. The corresponding diffraction pattern, Fig. 4.29*b*, contains only sharp spots.

Wilson (1949*a*) has given a theory of this effect based on the assumption that as ordering proceeds the gold atoms tend to move so that they are nowhere in contact with one another. He obtains the following expression for the intensity of reflection $H(u, v, w)$ at a point in reciprocal space having coordinates u, v, w:

$$H(u, v, w) = N(\text{Au}-\text{Cu})^2 . \frac{\alpha}{\alpha^2+(\pi u'')^2} . \frac{\alpha}{\alpha^2+(\pi v'')^2} . F(w''),$$

where N = number of unit cells in the crystal.

Au,Cu = atomic scattering factors for gold and copper atoms respectively.

α = probability of a mistake occurring at any point.

u'', v'' = coordinates in reciprocal space parallel to the two cubic axes chosen in the following way. The relp nearest to the reciprocal point defined by (u'', v'', w'') will have two indices both odd or both even and the third different in this respect from the other two. The axes to which the two odd or even indices refer are made those to which u'' and v'' refer. The third axis then defines the direction in which w'' is measured.

$F(w'')$ = function giving the variation of the intensity of scattering with distance from the relp in a regular crystal.

It will be seen that this theoretical expression gives a form to the diffuse region in reciprocal space which is disk-like. In the direction of the coordinate w'' the fall of intensity from the centre of the diffuse region is very rapid but in the direction of u'' and v'' it is relatively slow. The larger the value of α the greater must be the distance from the centre before the intensity falls to a given fraction of its value at the centre. The theory therefore predicts a result which is close to that which is observed.

4:5 **Diffuse scattering from age-hardening alloys**

A number of diffuse X-ray reflections are observed in single crystals of alloys such as Al-5 per cent. Cu, Al-20 per cent. Ag, Al-Mg-Si. At high temperature the solute atoms are in solid solution but on cooling to lower temperatures segregation of the solute atoms occurs. The manner in which this segregation occurs is different in the different alloys and we shall only deal with a few typical examples.

4:5.1 *Al-5 per cent. Cu*

One of the best studied age-hardening alloys is Al-5 per cent. Cu (Preston, 1938*a*, *b*; Guinier, 1938, 1942, 1949, 1950, 1952; Toman, 1955, 1957; Gerold, 1954, 1958; Doi, 1960). Above 520°C the copper atoms are distributed at random in the matrix of the aluminium atoms. After quenching and subsequent annealing at room temperature for some tens of hours or at 200°C for a few hours, a segregation occurs which results in certain diffuse X-ray reflections. The copper atoms are segregated into small platelets parallel to {100}, which are a few atoms thick and contain some tens or hundreds of atoms depending on the stage of development. The reciprocal equivalents of platelets parallel to planes of the form {100} are rods parallel to the axes ⟨100⟩. These diffuse rods are due both to the copper platelets themselves and also to the distortion of the aluminium matrix by these platelets. Fig. 4.30 shows an oscillation photograph of a single crystal of this alloy, the axis of oscillation being [001]. The diffuse streaks extending horizontally through the origin and on the right-hand side of the 200 spot are clearly visible. There are also diffuse streaks extending in a vertical direction above and below the central spot. This photograph is taken in crystal-reflected monochromatic radiation and the diffuse streaks are therefore entirely due to the atomic arrangement in the crystal. Following Preston (1938*a*, *b*), Guinier (1942) proposed a model to explain these diffuse streaks. He postulated platelets of copper atoms some tens or hundreds of Angstroms in diameter but only two atoms thick. The copper atoms are 11 per cent. smaller than the aluminium atoms they replace and stronger scatterers of X-rays (ratio of $f_{Cu}/f_{Al} \approx 2{\cdot}5$). Guinier found that a model represented by Fig. 4.31 satisfactorily explained the position and the intensity of the diffuse streaks. The relative spacing of the planes are shown on

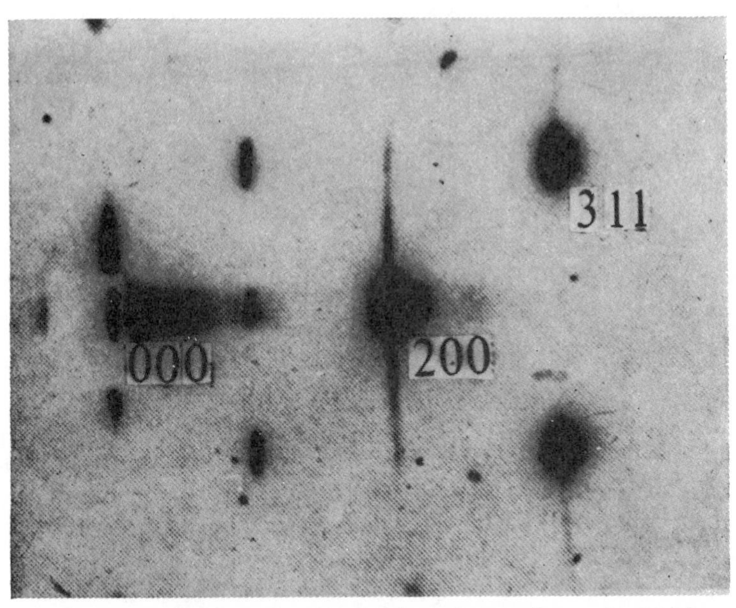

FIG. 4.30 Photograph showing horizontal streaks extending to the right from the 000 and 200 reflections of an age-hardened Al-5 per cent. Cu alloy. (After Guinier, 1945*b*).

Scattering power	Spacing
ρ	
	d
ρ	
	d
ρ	
	1.1 d
1.2 ρ	
	0.9 d
1.2 ρ	
	1.1 d
ρ	
	d
ρ	
	d
ρ	

FIG. 4.31 Diagram representing a model, due to Guinier (1942), showing the effect of a segregated platelet of copper in changing the scattering power, ρ, and the spacing, d, of the aluminium crystal.

the right-hand side and the relative scattering powers on the left-hand side. (This model has an inconsistency because the sum of the spacings of the disturbed atomic layers is 1·1+0·9+1·1 = 3·1 and is not an integral number.) Theoretical papers going thoroughly into the problem of diffraction by such platelets have been published by Toman (1955, 1957), Gerold (1954, 1958), and Doi (1960). The general picture originally put forward appears to be confirmed though different arguments concerning the exact thickness of the copper platelet and the allowance to be made for the disturbance of the lattice of the matrix by the platelet continue to be advanced.

The scattering close to 000 in Fig. 4.30 is due to the small-angle scattering of the thin platelets. It is the size and shape of the platelet, rather than the atomic distribution within it, which produces such streaks. From the extent and the intensity of these streaks Guinier estimates, on the assumption of a copper platelet two atoms thick, that for the alloy Al-4 per cent. Cu quenched and annealed for twenty-four hours at 100°C, the platelets are about 35 Å in diameter. The mean distance between such local concentrations of copper atoms is estimated to be 40 Å.

4:5.2 *Al-Ag, Al-Zn alloys*

In contrast with the Al-5 per cent. Cu alloy, which gives rise to precipitated platelets, the alloys of aluminium with silver or zinc give more or less spherical segregations of the solute atoms. This is shown by the small-angle scattering (Fig. 4.32*a*, *b*), which shows the contrast between the diffuse spikes of Al-5

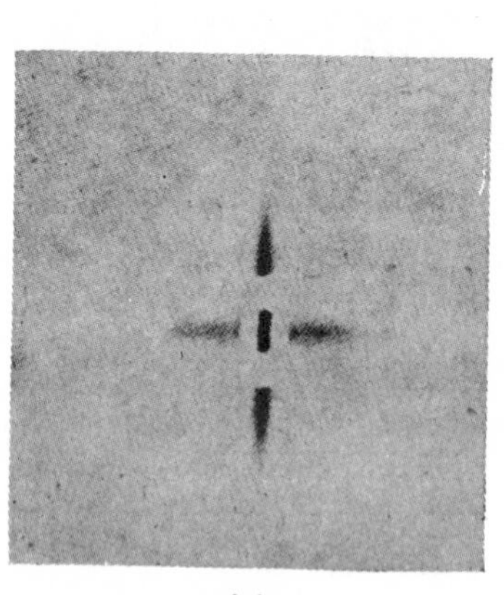

(*a*)

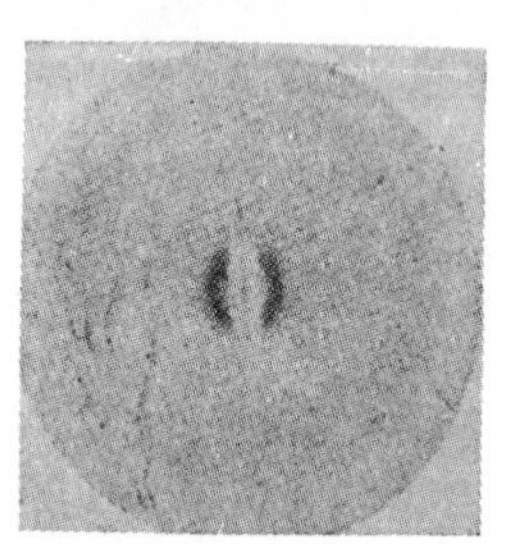

(*b*)

FIG. 4.32*a* Photograph showing the scattering near to the direct beam (of strictly monochromatic radiation) given by a single-crystal plate of an age-hardened Al-5 per cent. Cu alloy.

Fig. 4.32*b* The same as (*a*) but using an Al-Zn single crystal. (After Guinier, 1945*b*).

per cent. Cu (*a*) and the diffuse hollow sphere (*b*) surrounding the origin of the reciprocal lattice, in Al-20 per cent. Zn. The diameter of the hollow sphere depends on the mean separation of the local segregations of zinc atoms and the intensity depends on the number of zinc atoms in each segregation. For the above alloy the radii of the spheres were estimated to be 70 Å for the alloy aged at 15°C, 90 Å for that aged at 100°C, and 140 Å for that aged at 150°C.

In the aluminium-silver alloys the precipitated phase, $AlAg_2$, forms a close-packed hexagonal structure which has a preferred orientation relative to the aluminium matrix. The (00·1) plane of the precipitate is parallel to the (111) plane of the matrix. During the stages of annealing before the precipitate has finally separated out into a distinct phase various diffuse effects are observable.

(i) After annealing an alloy of Al-20 per cent. Ag at temperatures under 150°C cylindrical diffuse spikes occur parallel to the $\langle 111 \rangle$ directions of the matrix. The radius of the spikes is 1/10 to 1/20 of the distance between neighbouring point rows parallel to $\langle 111 \rangle$. The intensity varies little from point to point along these spikes.

(ii) Annealing for a long time between 100°C and 150°C enables diffuse disks to be observed. The disks are parallel to $\{100\}$ planes and their centres lie along $\langle 100 \rangle$ directions. The disks are $a^*/3$ and $2a^*/3$ apart from one another and from the normal relps of the matrix. The diameter of these disks is roughly a^* and their thickness $a^*/6$.

An intepretation of these diffuse effects can be derived as follows. The segregated silver atoms are grouped into more or less spherical domains as is shown by the small-angle scattering. The observation (i) indicates that the (111) planes of the local segregations are stratified so that they differ from the surrounding matrix though they are similar to one another. From the fact that in the precipitated phase, $AlAg_2$, the (00·1) plane of the hexagonal close-

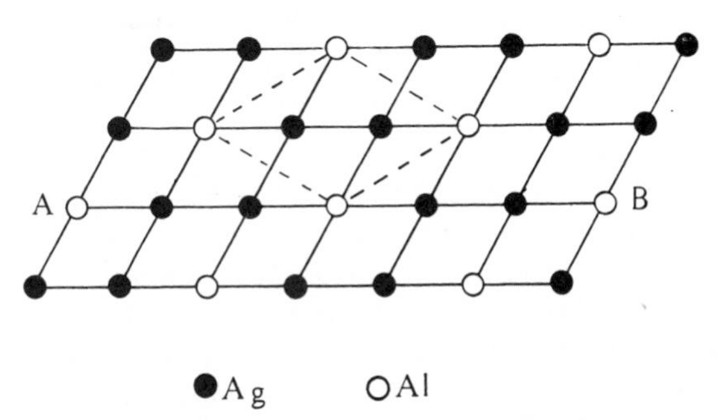

FIG. 4.33 Diagram showing the proposed distribution in a 111 plane of atoms in the alloy Al-Ag.

packing is parallel to the (111) plane of the cubic close-packed matrix, we may expect that on this plane the silver atoms will tend to be surrounded by a regular hexagon of aluminium atoms. In Fig. 4.33 is shown one plane of this arrangement. This would give the required ratio of aluminium to silver atoms and the proper coordination. The Fourier transform of such two-dimensional nets consists of diffuse spikes passing through the relps of the aluminium matrix in directions parallel to ⟨111⟩. This corresponds with the observation (i). The observation (ii) indicates that there is an approximation to a cubic superstructure with a cell side three times as long as that of the matrix. In Fig. 4.34 the plane of atoms shown in Fig. 4.33 is drawn on the cubic axes. It will be seen that a superstructure based on a cell of side $3a$ is consistent with this structure. The line AB represents the same set of atoms in Figs. 4.33 and 4.34. The complete structure is obtained by stacking not quite identical planes of atoms similar to that shown in Fig. 4.34 on top of one

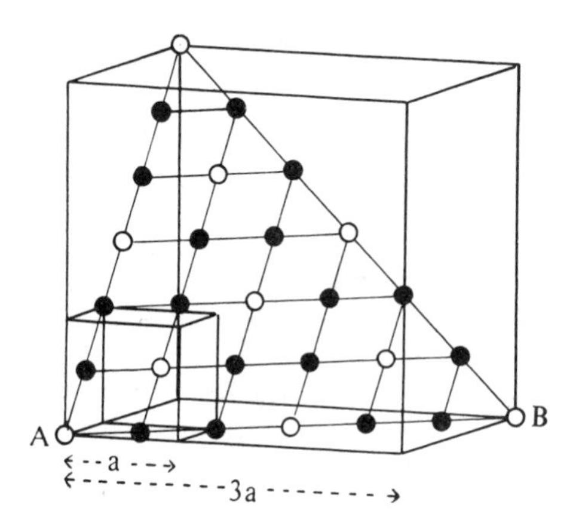

FIG. 4.34 Diagram showing how the distribution given in Fig. 4.33 fits into a cubic unit cell which has a cell side three times as long as that of the matrix.

another. Ignoring the difference between silver and aluminium, this whole arrangement is based on cubic close-packing. If there is local disorder brought about by the omission of an atom from one of the lattice sites, or the insertion of one of the wrong kind, the diffuse effects must be similar to those obtained in $AuCu_3$. The sharp relps are those corresponding to the face-centred cubic distribution lattice and have indices all odd or all even. The diffuse relps, associated with a superlattice of side $3a$ in the crystal structure, occur at distances of $a^*/3$ and $2a^*/3$ on either side of the sharp relps in the directions of the cell edges, where a^* refers to the reciprocal cell size of the matrix. As in $AuCu_3$ these diffuse relps are spread into disks having their planes normal to the axes ⟨100⟩. Thus Guinier explained his experimental results in terms of an ordered segregation leading towards the crystallization of a separate phase.

4:6 Diffuse X-ray reflections given by partially ordered alloys

4:6.1 *Long-range order*

Many alloys of two or more elements form a number of phases stable over certain ranges of temperature and composition. The condition is sometimes met in which, below a certain temperature, the atoms of an alloy arrange themselves on one or more lattices which are similar and similarly orientated. We have already frequently referred to the alloy Cu_3Au. In the state showing perfect long-range order the gold atoms are at the corners, and the copper atoms at the centres of the faces, of the unit cell. Under conditions of cooling which do not permit a perfect ordering of the two sorts of atoms, a partial approach to this state occurs. Bragg and Williams (1935) defined the degree of ordering by a parameter S, given by the equation

$$S = (p-r)/(1-r),$$

where p is the probability of finding in the alloy AB an A-type atom in the site appropriate to such an atom and r is the fraction of the whole number of atoms which are A-type atoms. Thus for perfect order p is unity and S is also unity; for complete disorder the probability of any atom being an A atom is equal to the proportion of A atoms present, i.e. r. In this case S is zero. Thus S is proportional to the amount of the crystal which is perfectly ordered. A face-centred lattice gives X-ray reflections for which the indices hkl are either all odd or all even. When gold atoms replace copper atoms at the corners of the unit cell the cell becomes primitive and indices hkl of any value may occur. The reflections for which the indices are all odd or all even are those to which all atoms contribute, so that the structure amplitude is proportional to $(3f_{Cu}+f_{Au})$. The other reflections have a structure amplitude proportional to the difference in the scattering power of the gold and copper atoms, i.e. $f_{Au}-f_{Cu}$. This can be seen by supposing that a primitive lattice consisting of atoms having a scattering power $f_{Au}-f_{Cu}$ is superimposed on a face-centred lattice of the same cell size consisting of copper atoms only. This double lattice would have the same scattering properties as the fully ordered Cu_3Au. To the reflections for which h, k, and l are mixed indices only the $(f_{Au}-f_{Cu})$ lattice would contribute and hence this gives the structure amplitude of these reflections.

In the disordered structure we note that S gives the fraction of the crystal which has the fully ordered arrangement. Hence we may write for the intensity I of the lines, known as superstructure lines,

$$I = (f_{Au}-f_{Cu})^2 . S^2.$$

By measuring the intensity of reflection corresponding to (100) and (110) lines of a powder pattern, the value of S may be determined. For the Alloy $CuAu_3$ this has been carried out by Batterman (1957).

4:6.2 *Short-range order*

A more difficult problem is presented by the ordering which occurs in certain alloys in the near neighbourhood of a given atom. If atoms of type B are dissolved in a lattice of atoms of type A several effects arise. The size and scattering power of atoms of type B will generally be different from those of type A. Moreover, atoms of type B may tend either to segregate together into small domains of a composition different from that of the matrix or to move as far apart from one another as possible. The differing sizes of the A and B atoms result in local deformations of the matrix. The difference in their scattering power gives rise to a certain contribution to the diffuse scattering. Superimposed on all these effects are the thermal vibrations of the atoms, which also contribute to the diffuse scattering.

A number of papers have been written with the object of founding a theory of the diffuse scattering that occurs in alloys owing to short range order. It is difficult to separate the various contributions to the experimentally measured diffuse scattering but in certain alloys this has been done fairly satisfactorily. The analysis tends to be rather heavy and here only an outline of the theory will be attempted. Many authors have contributed to this subject (Laue, 1943; Guinier, 1945*b*; Wasastjerna, 1947; Cowley, 1950*a*; Warren, Averbach, and Roberts, 1951; Chipman, 1956; Herbstein, Borie, and Averbach, 1956; Batterman, 1957; and Suoninen and Warren, 1958). The following treatment is based on a number of these papers but especially on those of Warren, Averbach, and Roberts (1951) and Batterman (1957).

A fundamental theorem, which was very clearly demonstrated by Cochran (1956), states that when certain atoms in a perfectly regular lattice are replaced by others of different scattering power or slightly different positions, the transform of the actual crystal, at points in reciprocal space not coinciding with normals relps, is given by the transform of the difference between the actual and the original perfect crystal. If we suppose that in an AB alloy the B atoms are precisely at the lattice points of the A atoms they replace, then the difference between the actual and the original crystal is a random distribution of $(B–A)$ atoms on some of the original lattice points. At all points removed from the normal relps the intensity given by such an arrangement is I, where

$$I = Nm_A . m_B(f_B - f_A)^2$$

and is spherically symmetrical about the origin (Zachariasen, 1945, p. 225). N is the number of atoms in the crystal. The proportions of A and B atoms are denoted m_A, m_B respectively. This equation corresponds to a completely random distribution of the $A–B$ scatterers. The fact that in our case the distribution is not random, but is restricted to lattice sites, makes no difference at points removed from the relps.

4:6.3 *Influence of differing atomic sizes*

If all the atoms are at lattice points the usual expression for the intensity of scattering is

$$I = N \sum_{m} \sum_{m'} f_m f_{m'} \exp[i\mathbf{K}^* . (\mathbf{r}_m - \mathbf{r}_{m'})], \tag{4.1}$$

where f_m, $f_{m'}$ are the atomic scattering factors for atoms of types m, m' respectively and $\mathbf{r}_m$, $\mathbf{r}_{m'}$ are the vectors joining them to the origin (Zachariasen, 1945, p. 225). N is the number of atoms in the crystal. To allow for differences in size we specify the distances between a given A atom and one in the ith shell round it as $r_{AA}{}^{i}$ and put

$$r_{AA^i} = r_i(1+\varepsilon_{AA^i}),$$

where r_i is the average distance between the central atom and all the neighbouring atoms of the ith shell. Similarly, we write

$$r_{AB^i} = r_i(1+\varepsilon_{AB^i}),$$

$$r_{BB^i} = r_i(1+\varepsilon_{BB^i}).$$

The short-range order is defined by a probability p_i which represents the probability of finding an A atom at a distance r_i from a B atom, and p_i' which represents the probability of finding a B atom at a distance r_i from an A atom. We may write

$$m_A p_i' = m_B p_i.$$

The quantity $(1-p_i')$ represents the probability of *not* finding a B atom at a distance r_i from an A atom. Since the alloy is composed of A and B atoms only this probability is also that of finding an A atom at a distance r_i from an A atom. Similarly, $(1-p_i)$ represents the probability of finding a B atom at a distance r_i from a B atom.

Equation (4.1) can now be expanded. The first term is one for which $m = m'$ and $(\mathbf{r}_m - \mathbf{r}_{m'}) = 0$, namely

$$N(m_A f_A^2 + m_B f_B^2).$$

For all succeeding terms $m \neq m'$ and $f_m, f_{m'}$ must be replaced by $f_A, f_B, p_i f_A$, $(1-p_i')f_A, p_i' f_B$ or $(1-p_i')f_B$ according to the selection of atoms. Thus if m and m' are both A atoms,

$$f_m = f_A, \qquad f_{m'} = (1-p_i')f_A.$$

Bearing in mind that

$$r_{AA^i} = r_i(1+\varepsilon_{AA^i}),$$

the distances between the A atoms are given by

$$(\mathbf{r}_m - \mathbf{r}_{m'})(1+\varepsilon_{AA^i}),$$

where $\mathbf{r}_m$, $\mathbf{r}_{m'}$ are the distances between the corresponding atoms in the undisturbed lattice. Thus for the term concerned only with A atoms we have the expression

$$m_A f_A(1-p_i')f_A \exp\{i\mathbf{K}^*.(\mathbf{r}_m-\mathbf{r}_{m'})(1+\varepsilon_{AA^i})\}. \tag{4.2}$$

Next we may take A atoms having B neighbours, for which the expression is

$$m_A f_A p_i' f_B \exp\{i\mathbf{K}^*.(\mathbf{r}_m-\mathbf{r}_{m'})(1+\varepsilon_{AB^i})\}. \tag{4.3}$$

When only B atoms having A neighbours are considered, we obtain

$$m_B f_B p_i f_A \exp\{i\mathbf{K}^*.(\mathbf{r}_m-\mathbf{r}_{m'})(1+\varepsilon_{AB^i})\}, \tag{4.4}$$

and, lastly, for B atoms having B neighbours the expression becomes

$$m_B f_B(1-p_i)f_B \exp\{i\mathbf{K}^*.(\mathbf{r}_m-\mathbf{r}_{m'})(1+\varepsilon_{BB^i})\}. \tag{4.5}$$

Since ε is small,

$$\exp\{i\mathbf{K}^*.(\mathbf{r}_m-\mathbf{r}_{m'})(1+\varepsilon)\} = \{1+i\mathbf{K}^*.(\mathbf{r}_m-\mathbf{r}_{m'})\varepsilon\}\exp\{i\mathbf{K}^*.(\mathbf{r}_m-\mathbf{r}_{m'})\}.$$

For brevity we shall write

$$i\mathbf{K}^*.(\mathbf{r}_m-\mathbf{r}_{m'}) = E.$$

Remembering that $m_A p_i' = m_B p_i$ and $m_A+m_B = 1$, we may make the following reductions:

$$\begin{aligned} m_A f_A^2+m_B f_B^2 &= m_A(m_A+m_B)f_A^2+m_B(m_A+m_B)f_B^2 \\ &= m_A^2 f_A^2+m_A m_B f_A^2+m_A m_B f_B^2+m_B^2 f_B^2 \\ &= m_A^2 f_A^2+2m_A m_B f_A f_B+m_B^2 f_B^2+m_A m_B f_B^2- \\ &\qquad -2m_A m_B f_A f_B+m_A m_B f_A^2 \\ &= (m_A f_A+m_B f_B)^2+m_A m_B(f_B-f_A)^2. \end{aligned}$$

The first term of equation (4.1), namely, $N(m_A f_A^2+m_B f_B^2)$, becomes

$$N(m_A f_A+m_B f_B)^2+N m_A m_B(f_B-f_A)^2 \quad \text{for } (\mathbf{r}_m-\mathbf{r}_{m'}) = 0. \tag{4.6}$$

The expressions (4.2) to (4.5) become, respectively,

$$\sum_m \sum_{m'} m_A(1-p_i')f_A^2(1+E.\varepsilon_{AA^i})\exp E, \tag{4.7}$$

$$\sum_m \sum_{m'} m_B p_i f_A f_B(1+E.\varepsilon_{AB^i})\exp E, \tag{4.8}$$

$$\sum_m \sum_{m'} m_B p_i f_A f_B(1+E.\varepsilon_{AB^i})\exp E, \tag{4.9}$$

$$\sum_m \sum_{m'} m_B(1-p_i)f_B^2(1+E.\varepsilon_{BB^i})\exp E. \tag{4.10}$$

When we take only terms not involving ε, the sum of the expressions (4.7) to (4.10) gives, when $(\mathbf{r}_m - \mathbf{r}_{m'}) \neq 0$,

$$\sum_m \sum_{m'} \{m_A f_A^2 - m_B p_i(f_A^2 - 2f_A f_B + f_B^2) + m_B f_B^2\} \exp E$$

$$= \sum_m \sum_{m'} \{(m_A f_A + m_B f_B)^2 + m_A m_B (f_B - f_A)^2 - m_B p_i (f_B - f_A)^2\} \exp E$$

$$= \sum_{\substack{m \\ m \neq m'}} \sum_{m'} \left[(m_A f_A + m_B f_B)^2 + \sum_m \sum_{m'} \left\{m_A m_B (f_B - f_A)^2 \left(1 - \frac{p_i}{m_A}\right)\right\}\right] \exp E. \quad (4.11)$$

Terms involving ε give the sum

$$\sum_m \sum_{m'} \{(m_A - m_B p_i) f_A^2 \varepsilon_{AA^i} + 2m_B p_i f_A f_B \varepsilon_{AB^i} + m_B(1 - p_i) f_B^2 \varepsilon_{BB^i}\} E.\exp E. \quad (4.12)$$

Finally, we obtain the intensity by adding the expressions (4.6), (4.11), and (4.12) to obtain

$$I = N m_A m_B (f_B - f_A)^2 +$$

$$+ \sum_m \sum_{m'} \left\{m_A m_B (f_B - f_A)^2 \left(1 - \frac{p_i}{m_A}\right)\right\} \exp E +$$

$$+ \sum_m \sum_{m'} \{(m_A - m_B p_i) f_A^2 \varepsilon_{AA^i} + 2m_B p_i f_A f_B \varepsilon_{AB^i} + m_B (1 - p_i) f_B^2 \varepsilon_{BB^i}\} \times E.\exp E +$$

$$+ \sum_m \sum_{m'} (m_A f_A + m_B f_B)^2 \exp E. \quad (4.13)$$

The first term in equation (4.13) gives the diffuse scattering, mentioned at the beginning of section 4:6.2, which is due to random distribution of the B atoms on the A lattice sites. The second term gives the diffuse scattering due to short-range order. The third term (ending in $E.\exp E$) gives the size-effect diffuse scattering which is due to the local distortions of the A lattice by the B atoms. The fourth term gives the normal sharp Bragg reflections.

The following substitutions permit a simplification of equation (4.13).

Put $$\eta = f_B / f_A, \qquad \alpha_i = \left(1 - \frac{p_i}{m_A}\right),$$

$$\beta_i = \left(\frac{1}{\eta - 1}\right)^2 \left\{\left(\frac{m_A}{m_B} + \alpha_i\right) \varepsilon_{AA^i} + 2(1 - \alpha_i)\eta \varepsilon_{AB^i} + \left(\frac{m_B}{m_A} + \alpha_i\right) \eta^2 \varepsilon_{BB^i}\right\}.$$

We may also replace the double sums by single sums multiplied by N since the effect of neighbours on short-range order is restricted to the first few shells.

If I_D is the diffuse intensity due to the short-range order and the size effect we may write, extending the summation over all j neighbours of the ith shell,

$$\frac{I_D}{N m_A m_B (f_B - f_A)^2} = \sum_{ij} \alpha_i \exp E + \sum_{ij} \beta_i E.\exp E. \quad (4.14)$$

If one atom is taken to be at the origin then E is simply the normal structure factor summation over the atoms in a unit cell, namely, $\sum 2\pi(hx+hy+lz)$. In this case h, k, l, are non-integral and x, y, z, are the coordinates of the atoms in the unit cell. The summation containing α_i's varies periodically with h, k, l, repeating in every succeeding reciprocal unit cell. The values of the α_i's can be found in a manner similar to that of Fourier synthesis in crystal structure determination. To do this, however, it is necessary to separate the contribution due to the size effect from that of short-range order. The size-effect term contains the product $E.\exp E$ and therefore increases steadily with distance from the origin of the reciprocal lattice. The following examples illustrate how the β_i-coefficients may be determined.

4:6.4 *Experimental measurements*

The experimental methods used in the study of the diffuse reflections from alloys are similar to those already described. Much use has been made of powders as well as of single crystals. The most readily interpretable data are obtained from single crystals. A study along a line $h00$ in reciprocal space (where h is a non-integral coordinate) gives a density distribution which can be analyzed into a short-range order and a size-effect contribution. This is shown in Fig. 4.35 due to Warren, Averbach, and Roberts (1951) for Cu_3Au.

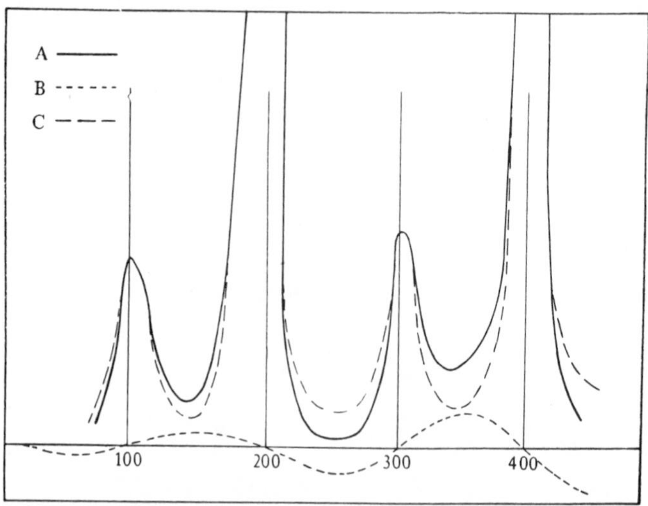

FIG. 4.35 Diagram showing the distribution of diffuse scattering power along the x^*-axis for a single crystal of Cu_3Au quenched from 500°C. Curve A gives measured values; curve B the calculated size-effect term due to nearest neighbours; and curve C, the difference between curves A and B. (After Warren, Averbach, and Roberts, 1951).

The experimental results are given by the line A, the short-range order by curve B, and the size-effect by curve C. The size-effect has been calculated using only the terms involving the twelve nearest neighbours, and the magnitude of β_1 is adjusted so as to make the difference between the ordinates for curves A and C give a background of the same height at points having coordinates 1·5, 2·5, and 3·5 along the x^*-axis. A two-dimensional survey of the diffuse reflection in the $(hk0)$ plane of the reciprocal lattice for the alloy $CuAu_3$ is shown in Fig. 4.36 due to Batterman (1957). The ordered structure of this

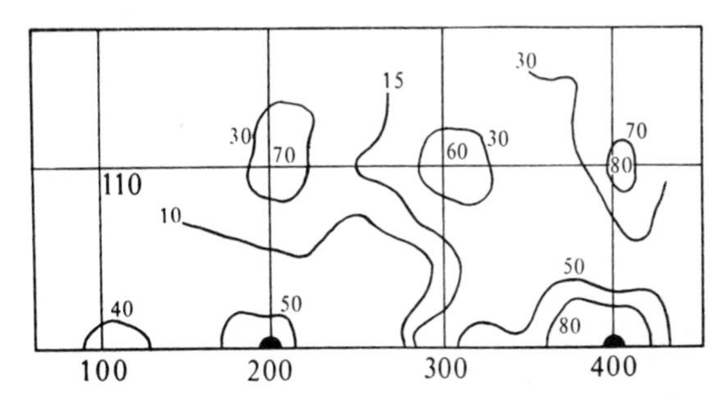

FIG. 4.36 Diagram showing the diffuse density distribution in the $hk0$ plane for a single crystal of $CuAu_3$ quenched from 285°C. (After Batterman, 1957).

alloy is the complementary one to that of Cu_3Au, i.e. the copper atoms are at the centres of the faces of the unit cell and the gold atoms are at the corners. The usual Bragg reflections in the figure are thus 200 and 400. The diffuse peaks about 100, 210, 300, 310, and 410 are due to the combination of various effects. The separation of the contributions from temperature diffuse scattering, Compton effect, short-range order, size effect, and still others which may be present, is difficult and the numerical results obtained are subject to large possible errors. The observed short-range order coefficients for Cu_3Au and $CuAu_3$ are shown in Table 4.1. The measurements on Cu_3Au were made at 405°C while those on $CuAu_3$ were made at room temperature on the specimen quenched from 285°C. There is a general similarity between the α_i's for the two alloys but a striking difference in the shape of the intensity distribution about the relps. In Fig. 4.36 it can be seen that the contours around 210 are elongated towards 200. Those around 300 and 410 are elongated towards 400. The diffuse cloud is thus egg-shaped around 200, 300, and 410, the long axis of the egg joining the strongly reflecting relps. This is in contrast with Cu_3Au where the corresponding diffuse cloud is disk-shaped. Further analysis of curves showing the intensity of diffuse reflection over an area of reciprocal space enables the size-effect coefficients to be determined. Knowing the α_i's, the atomic displacements corresponding to the distortion produced by large atoms in a matrix of smaller atoms can be approximately evaluated.

Cochran and Kartha (1956*b*) have evaluated the size-effect contribution due to a single interstitial atom placed at the body-centre of a face-centred

TABLE 4.1

Observed short-range order coefficients (α_i) for Cu_3Au and $CvAu_3$

Coordination shell (i)	α_i			
	Cu_3Au 405°C	$CuAu_3$ 285°C	*Perfect order*	*Complete disorder*
1	−0·15	−0·10	−0·33	0·00
2	+0·19	+0·24	+1·00	0·00
3	+0·01	−0·09	−0·33	0·00
4	+0·10	+0·11	+1·00	0·00

cubic unit cell. The effect falls into two parts, namely, that due to all displacements of atoms lying within a shell of radius 4·7 Å and that due to displacements of all atoms outside this shell. The contributions due to these two causes differ from one another considerably as K^* varies and, although the total effect is similar to that deduced in the experiments of Warren, Averbach, and Roberts (1951), it is probable that the theory given above is in need of revision to take account of the differing contributions from the very near and the further off neighbours of any one foreign atom.

V

DIFFUSE X-RAY REFLECTIONS FROM MOLECULAR CRYSTALS

5:1 **Introduction**

Whereas the diffuse X-ray scattering from simple inorganic structures, such as those of aluminium or potassium chloride, is confined to regions close to reciprocal lattice points, the scattering from molecular crystals is distributed over relatively large volumes of reciprocal space. Further, these diffuse volumes need bear no relation to the presence of relps corresponding to strong Bragg reflections; in fact it often happens that a diffuse cloud includes relps which have zero intensity of reflection. However, around relps there may be regions of strong diffuse reflections which are similar to the corresponding effects observed in simple inorganic structures. Apart from these two main features of the diffuse scattering there are often a number of rods or plates in reciprocal space which frequently extend between the normal relps. Some of these may be due to structural defects but others are of thermal origin and are as yet unexplained.

The interpretation of these observations has been, and still is, the subject of much discussion. It is probable that the diffuse clouds far removed from normal relps are due to molecular vibrations and librations. It is also likely that the scattering near to relps is related to the elastic constants in the same way as it is with simple inorganic compounds. The diffuse clouds are certainly related to the molecular shape and orientation and one of the applications of study of diffuse reflections is to the determination of the orientation of chain-like or plate-like molecules relative to the axes of the unit cell. A complete theory would undoubtedly involve the lattice dynamics of the crystal and when such a theory becomes established more use will be made of the quantitative data which are now available from the study of the diffuse patterns.

5:2 **Theoretical**

5:2.1 *The Difference Fourier Transform*

If a lattice composed of points which are in random independent thermal motions, scatters X-rays, then the intensity, $\bar{J}$, of scattering is given by James (1948) according to the expression

$$\bar{J} = \frac{|\Phi_0|^2}{R^2} N(1-e^{-2M}) + J_0 e^{-2M}, \tag{5.1}$$

where Φ_0 is the amplitude of the scattered wave at unit distance from the scattering point in the direction of the scattered wave normal,
R is the distance from the scattering point at which $\bar{J}$ is specified,
N is the number of lattice points in the crystal,
$-2M$ is the exponent of the Debye temperature factor and

$$M = 8\pi^2\overline{u_s^2}(\sin^2\theta)/\lambda^2,$$

$\overline{u_s^2}$ is the mean squared amplitude of vibration of the lattice points,
J_0 is the intensity of scattering given by the lattice, in the same direction as for $\bar{J}$ at a temperature of absolute zero.

The second term in this expression gives the (Bragg-Laue) intensity of scattering at the reciprocal lattice points and is not of interest here. The first term gives the scattering from all elements of reciprocal space except at the lattice points. We know that Φ_0 is proportional to the atomic structure factor, f_0, in a lattice of identical atoms, and this decreases continuously with increasing θ. N is the number of lattice points and is proportional to the volume of the crystal. The expression $(1-\exp(-2M))$ rises steadily with θ from zero at $\theta = 0$ to a maximum at $\theta = \pi/2$. Thus the product of $|\Phi_0|^2$ and $(1-\exp(-2M))$ gives a curve with a broad maximum at values of θ lying between 0 and $\pi/2$. Thus, this first term in equation (5.1) gives the diffuse scattering to be expected from point-like atoms. Finally, we may write for the intensity $I(\mathbf{H}^*)$ of the diffuse scattering far from reciprocal points

$$I(\mathbf{H}^*) = |f_0|^2 - |f_T|^2 \tag{5.2}$$

since

$$f_T = f_0 e^{-2M}.$$

This result, applying to a crystal composed of point-like atoms arranged on a primitive lattice, can be generalized. Cochran (1956) has shown that the diffuse scattering from a crystal with defects has an intensity equal to the difference of the square of the Fourier transforms of the perfect and the imperfect crystal, except at the reciprocal lattice points. In the present case the Fourier transform of the perfect crystal is simply $(F_{\mathbf{H}^*})_0$, the structure amplitude at the point $\mathbf{H}^*$ in reciprocal space at absolute zero. Also, the Fourier transform of the crystal with vibrating atoms is the structure amplitude at a temperature T, namely $(F_{\mathbf{H}^*})_T$. Thus the diffuse intensity at the point H^* is given by

$$I(\mathbf{H}^*) = (F_{\mathbf{H}^*})_0^2 - (F_{\mathbf{H}^*})_T^2. \tag{5.3}$$

The quantity $I(\mathbf{H}^*)$ has been called 'Difference Fourier Transform' (abbreviated to DFT) by Amorós, Canut, and Bujosa (1957) and was first used in X-ray diffuse scattering in the interpretation of their experimental results (see also Amorós and Canut, 1958*a*).

The physical basis for this result may be illustrated by the two-dimensional diagrams of Figs. 1.2*a* and 1.3*a*. Fig. 1.2*a* represents a projection of a monoclinic lattice on the (010) plane; the corresponding diffractogram has sharp spots. The points of the lattice in Fig. 1.3*a* are randomly displaced from their proper positions. This pattern when used as a mask in the optical diffractometer produces the pattern shown in Fig. 1.3*b*. The intensity of the diffuse scattering relative to that of the normal Bragg reflections increases with distance from the centre. With such large displacements from the correct positions only the first two spectra of the perfect arrangement are visible in Fig. 1.3*b*. Fig. 5.1*a* represents a single naphthalene molecule and Fig. 5.1*b* gives the

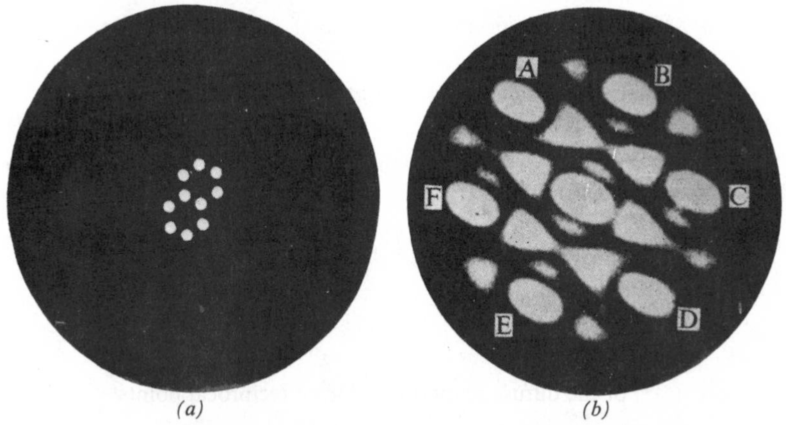

(*a*) (*b*)

FIG. 5.1*a* Mask representing a single molecule of naphthalene, having its plane parallel to the paper and in the same orientation as in Fig. 1.2*a*.

FIG. 5.1*b* Optical diffraction pattern given by the mask of Fig. 5.1*a*. (After Lipson and Taylor, 1951).

corresponding optical diffractogram. By comparing Figs. 1.3*b* and 5.1*b* it will be seen that the diffuse scattering from the much disturbed lattice is similar to that given by a single molecule. This optical illustration gives only an incomplete indication of the basis of the molecular diffuse scattering. The molecules are, in fact, not all parallel since thermal vibrations cause rotations as well as translations of the centres of the molecules. The optical model of Fig. 1.3*a* also takes no account of the internal vibrations of the molecules. In spite of these limitations the optical model shows how the thermal vibration of the centres of the molecules gives rise to a more or less continuous background which increases in intensity with distance from the centre.

5:2.2 *The calculation of Difference Fourier Transforms in molecular crystals*

The expression for the Fourier transform, $T_{(\mathbf{H}^*)}$, of any crystal is given by the well-known formula

$$T_{\mathbf{H}^*} = \sum_i f_i e^{-2\pi i(\mathbf{H}^* . \mathbf{r}_i - \nu t)}.$$

At absolute zero the f_i's are the values of the atomic scattering factors calculated by various authors and given in the standard tables. At temperatures above absolute zero the atoms are vibrating and in molecular crystals the amplitude of vibration may be anisotropic. Thus the amount of vibration of any atom in a planar molecule may be appreciably different according to whether the vibration occurs in the plane or normal to it. Cruickshank (1956, 1957) has shown how the calculations of the f_i's may be carried out for anthracene.

The treatment, which assumes an isotropic vibration of the atoms in the crystal, applies a term e^{-M} to the f-value at absolute zero of each atom in the usual structure amplitude formula. The value of M is given by James (1948) as follows:

$$M = 8\pi^2 \overline{u_s^2}(\sin^2 \theta)/\lambda^2,$$

where $\overline{u_s^2}$ is the mean value of the square of the amplitude of vibration of the atoms. When the thermal vibrations are anisotropic we may (following Cruickshank) write

$$\overline{u_s^2} = \sum_{i=1}^{3} \sum_{j=1}^{3} U_{ij} l_i l_j,$$

where U_{ij} is a symmetric second order tensor and the l's are the direction cosines of the direction of vibration considered. The values of U_{ij} were found by successive refinements of the parameters of the crystal structures with a technique employing digital computers. The six components of U_{ij} were determined for each of the atoms in the anthracene molecule.

Up to this point the analysis is simply a refinement of crystal structure analysis providing an improved calculation of $(F_{\mathbf{H}*})_T$. A result emerges, however, which has great importance for the theory of the diffuse scattering from molecular crystals. It can be shown that the internal vibrations of atoms within a molecule such as anthracene are small compared with the bodily movement of the whole molecule. Cruickshank (1956) therefore assumes in his analysis of the molecular movement that it may be regarded as a combination of the vibration and libration of the rigid molecule. The tensor U_{ij} can be resolved, using this assumption, into two tensors, denoted T_{ij}, ω_{ij}, the components of which define the vibration and libration respectively. Each of these tensors is symmetric and the diagonal elements are much greater than the off-diagonal elements. The vibration and libration tensors may be represented by triaxial ellipsoids and because of the small magnitudes of the off-diagonal components the principal axes of these ellipsoids almost coincide with the directions of the length, breadth, and thickness of the molecule.

The data for the libration of the molecule can be used to calculate the average frequencies of the rotational branches of the normal lattice vibrations. The values so calculated are 81, 65, and 43 cm^{-1} for the branches corresponding to oscillations about each of the molecular axes. Raman shifts in the light scattered from anthracene have been shown by Fruhling (1951) to yield the corresponding values 120 ± 8, 68 ± 5, 48 ± 4 cm^{-1}. In view of the assumptions made and the possible errors, this agreement is considered satisfactory.

5:2.3 *The scattering by thermal waves*

The theory based on the Difference Fourier Transform appears to differ from the theory used in Chapter II which is based on the reflections from elastic waves generated by thermal motion. But this is not the case as is shown by the theory for molecular crystals based on thermal waves put forward by Hoppe (1956*a*). The basic conceptions, which are the same as those used in Chapter II, apply strictly to a unit cell containing only one atom. For regions of reciprocal space close to relps Hoppe shows that a molecular crystal can be treated as though the whole molecule were replaced by a single point-like scatterer. Thus the elastic constants should be obtainable from the intensity of the diffuse scattering given by points near to relps. Far from relps the approximations made in the theory become more serious but the main result emerges that the scattering should be the same as that given by a point-like scatterer multiplied by the Fourier transform of the molecule or molecules in the unit cell. This result is the same as that given by equation (5.3). Hoppe's theory is quite general but it is not possible to apply it in detail because insufficient physical data are available. A problem of fundamental importance is the extent to which the molecules may be regarded as independent vibrating units. When scattering is produced by elastic waves the wavelength extends over hundreds of molecules and their relative movements imply a coupling between the vibration of any one molecule and its neighbours. When scattering occurs from regions roughly mid-way between relps this coupling need not be so important in determining the intensity of scattering. In fact the experimental results support the view that in this region the scattering is produced by molecules vibrating independently of one another. A theory has yet to be devised which will bridge the gap between the thermal wave treatment of Hoppe and the Difference Fourier Transform treatment.

5:3 **Experimental observations**

5:3.1 *Diffuse scattering from spherical molecules*

Hexamethylene tetramine, $C_6N_4H_{12}$, is a cubic structure with almost spherical molecules arranged in cubic close-packing. The first study of its diffuse reflection was carried out by Ramachandran and Wooster (1951*b*), who found the elastic constants by the method described in Chapter II, already established for inorganic crystals. They showed that the intensity of diffuse scattering of hexamethylene tetramine close to the 440 and 220 relps could be consistently explained in terms of the Waller-Faxen theory given in Chapter II. The elastic constants were determined as follows: $c_{11} = 1{\cdot}5$, $c_{44} = 0{\cdot}7$,

$c_{12} = 0{\cdot}3\times10^{11}$ dyne/cm^2, and the elastic ratios were $\chi_1 = c_{12}/c_{11} = 0{\cdot}21$, $\chi_2 = c_{44}/c_{11} = 0{\cdot}44$. The elastic constants of this substance have been measured by Haussühl (1958) by an ultrasonic method and he found the values $c_{11} = 1{\cdot}643 \pm 0{\cdot}005$, $c_{12} = 0{\cdot}433 \pm 0{\cdot}011$, $c_{44} = 0{\cdot}515 \pm 0{\cdot}005\times10^{11}$ dyne/cm^2. The corresponding elastic ratios were $\chi_1 = c_{12}/c_{11} = 0{\cdot}26$, $\chi_2 = c_{44}/c_{11} = 0{\cdot}31$. These results show that the X-ray method, although inaccurate, nevertheless gives a result based on an approximately correct theory. If this is so, the localized diffuse regions are due to the thermal waves and are subject to the laws discussed in Chapter II.

The next study of hexamethylene tetramine was due to Ahmed (1952), who plotted from Laue photographs the diffuse regions in a plane parallel to (100) passing through the origin. He found relatively dense localized regions near to the relps (which were the regions studied by Ramachandran and Wooster (1951*b*)) and also broad non-localized clouds round the relps 020, 011, and 022. Ahmed emphasized the existence of non-radial streaks in his photographs but gave no detailed explanation of the broad clouds or of the streaks. Amorós, Canut, Annaka, and de Acha (1958) also used a succession of Laue photographs and built up from them a more complete picture of the diffuse scattering. Fig. 5.2 shows a typical Laue photograph in which the broad

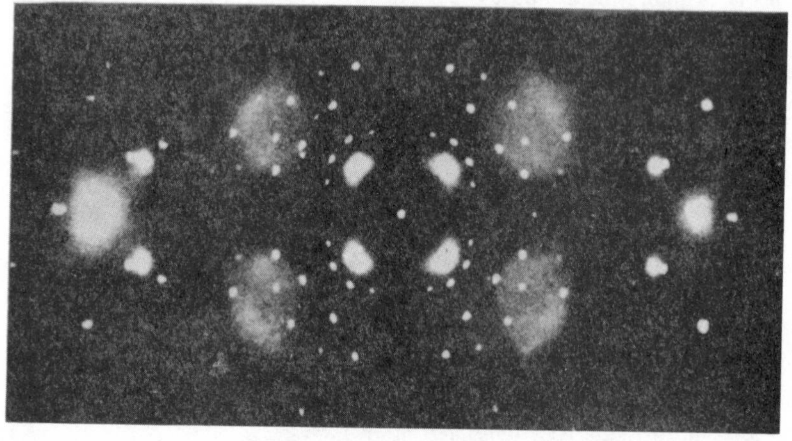

FIG. 5.2 Laue photograph of hexamethylene tetramine, showing broad regions of diffuse reflection. Axis [001] vertical. (After Amorós, Canut, Annaka, and de Acha, 1958).

regions may be clearly seen. The diffuse clouds extend continuously over large regions of reciprocal space but they are densest around the relps 110, 200, 222, 440, and 800. Along the [111] direction the cloud extends from 111 to 333 and around 222 the eight clouds round similar relps join one another. The DFT was calculated for the planes containing reciprocal points $hk0$, $hk\frac{1}{2}$, i.e. on reciprocal planes normal to the axis [001] and intersecting it at the origin and

FIG. 5.3*a* Experimentally plotted diffuse scattering on a section $hk\frac{1}{2}$ of hexamethylene tetramine.

FIG. 5.3*b* The calculated DFT corresponding to the observed distribution of Fig. 5.3*a*. (After Amorós, Canut, Annaka, and de Acha, 1858).

at a distance of $a^*/2$ from the origin respectively. The comparison between experimental and calculated distributions in the section $hk\frac{1}{2}$ is shown in Fig. 5.3*a*, *b*. It will be seen that the regions *A*, *B*, *C* correspond in the two figures. The region *D* was not plotted experimentally because of difficulty in distinguishing the general background from the effect of the diffuse cloud. Thus the *DFT* gives a reasonably satisfactory explanation of the observations. The diffuse scattering extends continuously through many relps, including some which have zero intensity of Bragg reflection. Formally this implies that the space group is not the same for the crystal structure with its molecules vibrating as it is for the structure with its molecules at rest.

5:3.2 *Diffuse scattering from chain-like molecules*

The diffuse scattering from certain chain structures has been investigated by a number of authors, including Lonsdale (1942*b*), Hoppe and Baumgärtner (1957), and Amorós and Canut (1957, 1958*b*). From the various Laue, Weissenberg, and other photographs the reciprocal lattice distribution corresponding to long parallel molecules can be obtained. For the fatty acids having between three and six carbon atoms in the aliphatic chain, there are two principal features of the diffuse distribution in reciprocal space. The first consists of a set of parallel continuous sheets which are perpendicular to the length of the molecules. The spacing of the sheets is inversely proportional to the length of the molecule and those sheets are densest which are at a distance from the origin corresponding to repeat distances along the molecular chains. For instance, in the aliphatic acids the repeat distance of 2·5 Å between alternate carbon atoms along the chain gives an enhancement of the diffuse layer whenever its distance from the origin corresponds to 2·5, 2·5/2, 2·5/3 Å, etc. Fig. 5.4

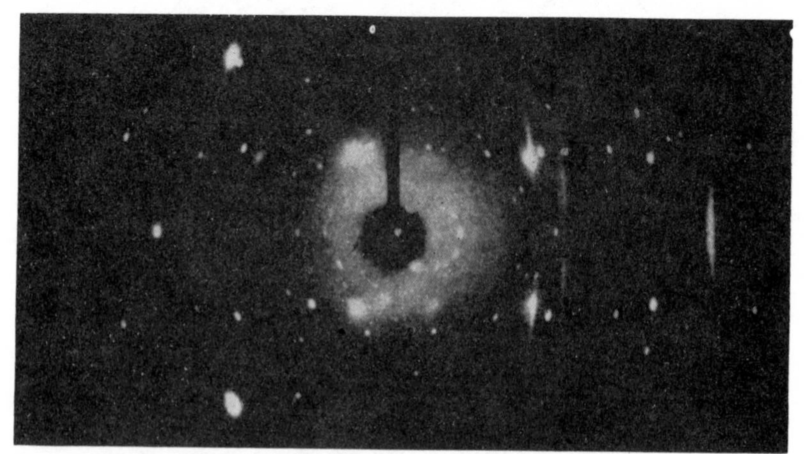

FIG. 5.4 Laue photograph of suberic acid, $COO(CH_2)_6$ COOH, with the molecular chains horizontal and diffuse sheets vertical. (After Amorós and Canut, 1958*b*).

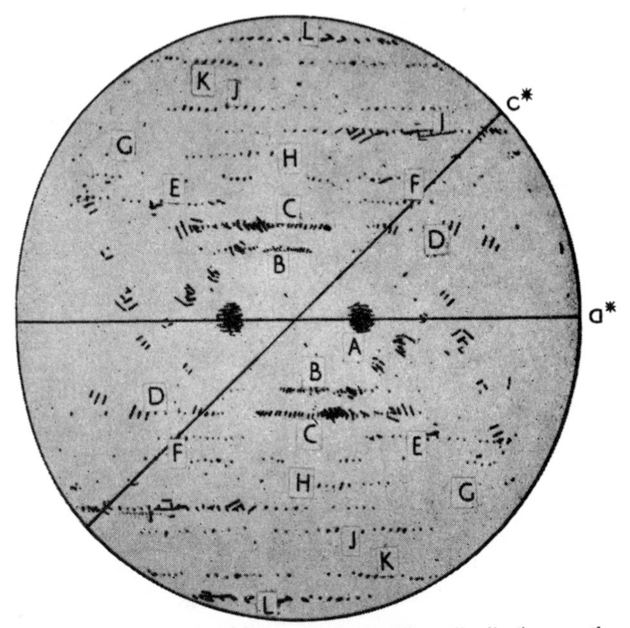

FIG. 5.5*a* Experimentally determined diffuse distribution on the section *h*0*l* of adipic acid, $COOH(CH_2)_4COOH$.

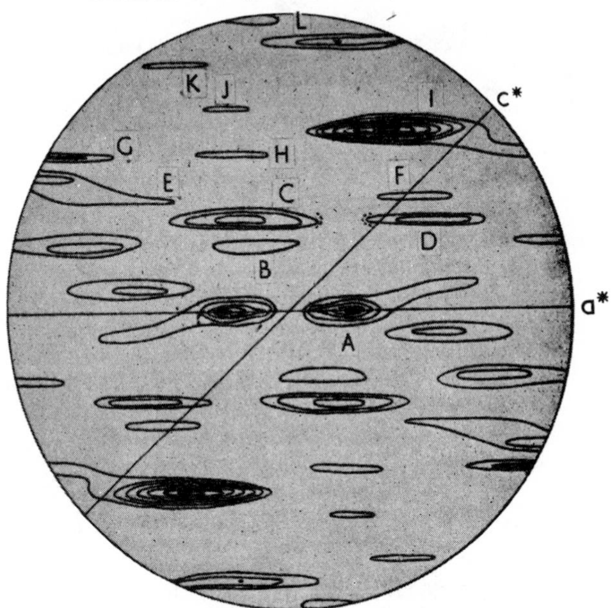

FIG. 5.5*b* Calculated DFT corresponding to the section given in Fig. 5.5*a*. (After Amorós and Canut, 1958*b*).

is a Laue photograph of suberic acid (Amorós and Canut, 1958*b*) which shows the diffuse row lines corresponding to the successive sheets perpendicular to the length of the chains. The diffuse sheets are bounded roughly by a cone of semi-angle 45° having its apex at the origin and its axis along the length of the molecules. The distribution of intensity within the sheets is not subject to the same space-group limitations as apply to the normal Bragg reflections. The second principal feature consists of intense diffuse clouds round certain relps which correspond to planes parallel to the length of the molecules. The shape of these intense clouds is roughly ellipsoidal and their greatest diameter is not usually more than twice as long as their minimum diameter. Fig. 5.5*a,b* shows the experimental and DFT distributions corresponding to the $[010]^0$ section of the reciprocal lattice of adipic acid according to Amorós and Canut (1958*b*). There is general agreement between the calculated DFT and the qualitative observations of the parallel diffuse sheets. It is safe to conclude that this feature of the diffuse pattern is adequately interpreted as the Fourier transform of the molecule modified by the thermal factor $(1-\exp(-2M))$. The intense diffuse spots mainly concentrated on the equator are not, however, satisfactorily accounted for by the DFT.

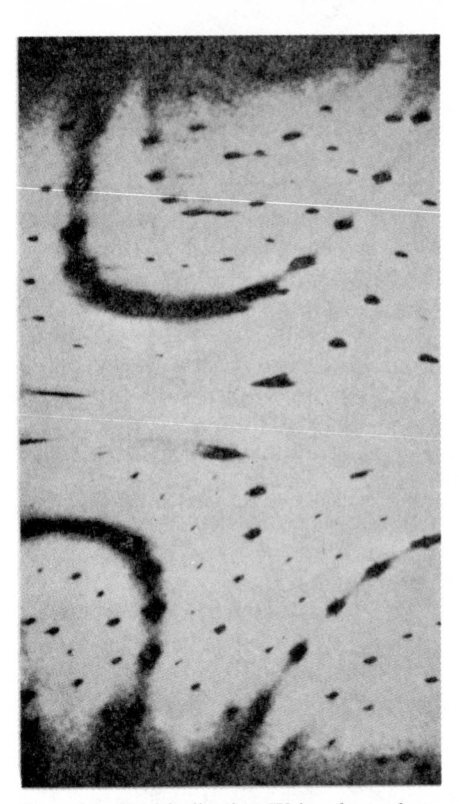

FIG. 5.6 Equi-inclination Weissenberg photograph of cyanurictrichloride, giving the section *h*5*l* of reciprocal space. (After Hoppe, Lenné, and Morandi, 1957).

5:3.3 *Diffuse scattering from structures containing flat molecules*

The diffuse patterns of substances containing flat molecules which are orientated parallel to one another have characteristic features. In reciprocal space diffuse rods extend in a direction normal to the plane of the molecule. If a Weissenberg photograph is taken with the crystal rotating about an axis perpendicular to the diffuse rod, and an appropriate reciprocal layer is chosen to include the rod, then the appearance shown in Fig. 5.6 is obtained (Hoppe, Lenné, and Morandi, 1957). The flat molecules of cyanurictrichloride, $C_3N_3Cl_3$, are parallel to the [010] axis about which the crystal was rotating during the taking of the photograph. The indices of the spots along the diffuse rods have l constant, showing that the rods are parallel to c^*, i.e. normal to the plane of the molecule. A precession photograph can also reveal the existence of these rods. Fig. 5.7 is such a photograph of Phyllochlorinester

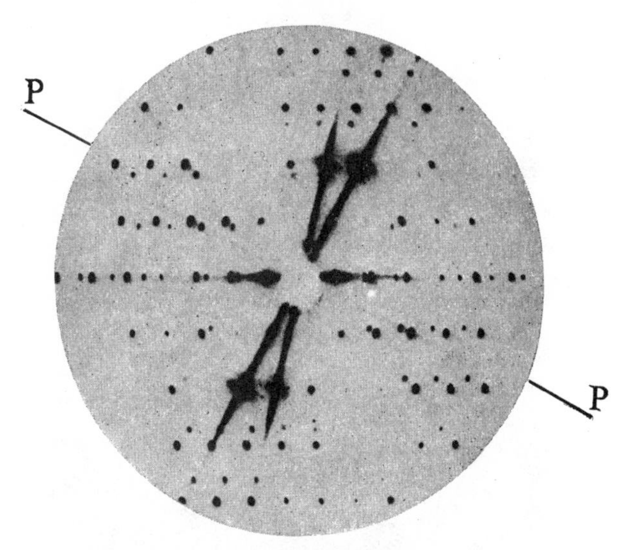

FIG. 5.7 Precession photograph of Phyllochlorinester, giving the $h0l$ section of reciprocal space. The plane of the molecule is parallel to PP and also perpendicular to the plane of the paper. (After Hoppe, 1957).

(Hoppe, 1957) taken about the [010] axis, and the indices of the spots are of the type $h0l$. Two dense rods extending from the origin through about three lattice points can be seen and these are perpendicular to the plane of the molecule, which intersects the paper in the line PP. The great enlargement of some of the spots along these reciprocal rods is a characteristic feature of such photographs. The diffuse scattering of anthraquinone (Hoppe, 1956*b*) shows similar features.

Anthracene

Another example of structures with flat molecules is provided by anthracene, $C_{14}H_{10}$. Unlike the previous examples the molecules have two orientations with respect to the edges of the unit cell. The long axes of the two molecules are inclined at an angle of a few degrees to the c-axis of the monoclinic cell. The axis which lies in the plane of the molecule and is perpendicular to its length, is inclined at about 29° to the *b*-axis and thus the two corresponding axes of molecules related by a plane of symmetry are inclined at about 58° to one another. Although Charlesby, Finch, and Wilman (1939) used electrons and not X-rays to study the diffuse scattering by anthracene their work is so close to our present discussion that it must be mentioned. A number of photographs are given similar to that shown in Fig. 5.8, together with diagrams of the orientation of the molecules relative to the X-ray beam. The photograph of Fig. 5.8 was taken with the electron beam almost normal

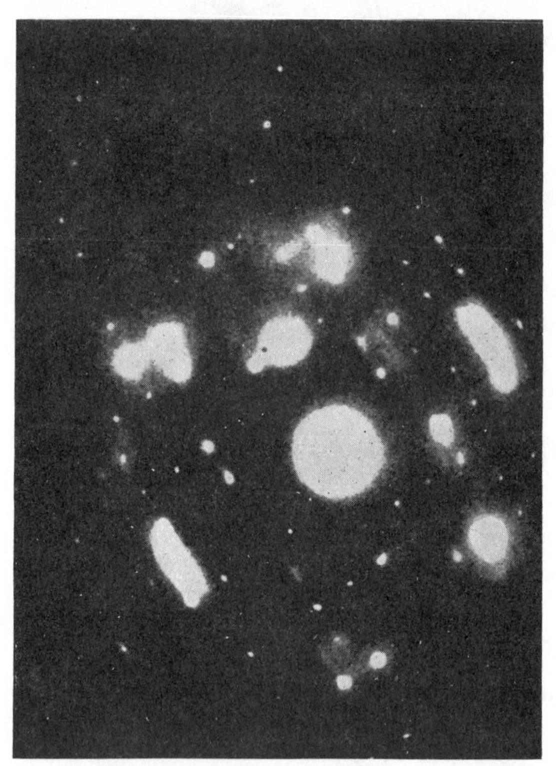

FIG. 5.8 Electron diffraction photographs of anthracene, showing hexagonally arranged diffuse rods. (After Charlesby, Finch, and Wilman, 1939).

to the long axis of both molecules, namely, along the axis [201]. Six regions of diffuse reflection will be noticed round the central spot. These correspond to the six rods of the molecular Fourier Transform which are perpendicular to the plane of a molecule. Amorós, Canut, Annaka, and de Acha (1958) have used X-rays and, by means of a succession of Laue photographs, they have traced the regions of diffuse scattering in a qualitative manner, and plotted sections of reciprocal space corresponding to the points $h0l$; $h\frac{1}{2}l$; $h1l$; $h, 3/2, l$; and $h2l$. The photographs show relatively broad rods directed along the normals to the planes of the molecules. These results are in agreement with those obtained by electrons mentioned above. Annaka and Amorós (1960) have continued this study also using a diffractometer. Sándor (1960) has carried out observations using Weissenberg photographs instead of a succession of Laue photographs. The Weissenberg photographs, using filtered Cu $K\alpha$ radiation corresponding to sections $(h\frac{1}{2}l)$ and $(h, 3/2, l)$ (Fig. 5.9), were studied using a recording microdensitometer and plots were made of contours of given density. Before making such density measurements consideration has to be given to the corrections which should be applied. We shall take these in turn.

5:3.4 *Corrections to measured intensities*

5:3.4.1 *Geometrical correction*

Consider an area $PQRS$ (Fig. 5.10*a*) of reciprocal space in the equatorial plane rotating at a uniform angular velocity ω about an axis passing through 0 normal to the paper. The semi-circle OSX represents the reflecting sphere. The Weissenberg photograph is represented in Fig. 5.10*b* and the area $pqrs$ is supposed to correspond with the area $PQRS$ in Fig. 5.10*a*. According to the usual relations applying to an equatorial Weissenberg photograph,

$$ps = k\delta\phi,$$

where k is the linear traverse of camera per unit angle of rotation of the crystal and $\delta\phi$ is the angle subtended by PS at the origin. If PQ corresponds to a change in the Bragg angle of $\delta\theta$ then the perpendicular distance between qr and ps is given by $2r\,\delta\theta$, where r is the radius of the camera. Thus the area, A, of the film $pqrs$ corresponding to the area $PQRS$ in the reciprocal lattice is given by

$$A = 2kr\,\delta\phi\,\delta\theta.$$

If the sides of the square $PQRS$ are of length δx^*,

$$\delta\phi = \delta x^*/x^*,$$

and since

$$x^* = 2\sin\theta,$$

$$\delta\theta = \delta x^*/2\cos\theta$$

and

$$A = kr(\delta x^*)^2/\sin 2\theta.$$

The intensity of blackening of the film per unit area of the reciprocal lattice for a given value of θ is inversely proportional to A and directly proportional to $\sin 2\theta$.

This, however, is not the only geometrical factor which must be considered; there is also the correction named after Lorentz. If any point such as P in

Fig. 5.10*a* is treated, not as infinitesimally small, but as a small spherical domain, then the time it takes for such an element of volume to pass through the reflecting sphere varies with θ. If ω is the angular velocity of rotation of the crystal (and of the reciprocal lattice), the linear velocity of P along PS is ωx^*. The component of this velocity along SC, a radius of the reflecting

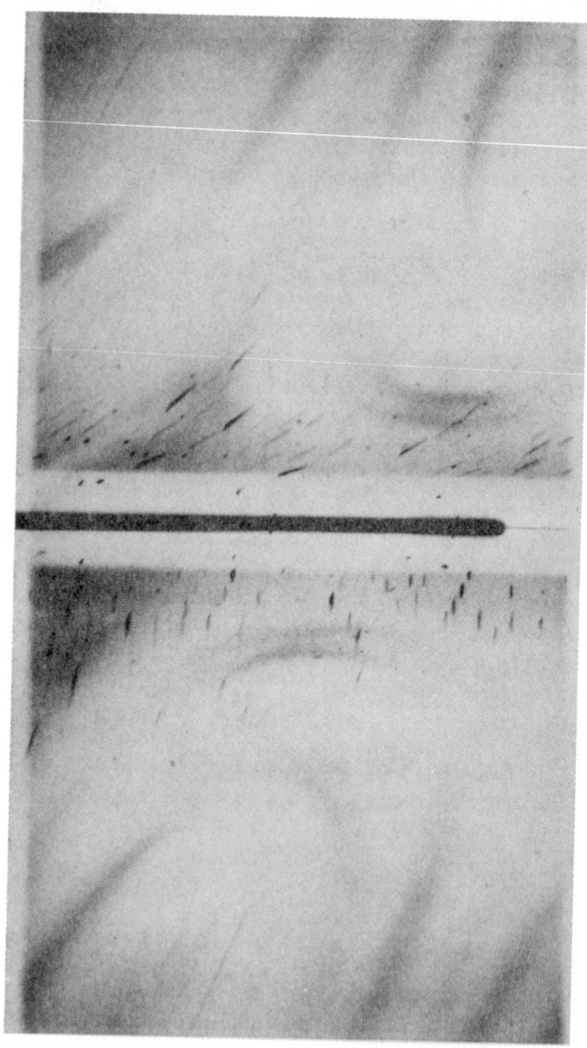

FIG. 5.9 Equi-inclination Weissenberg photograph of anthracene, $C_{14}H_{10}$, giving the reciprocal section corresponding to h, 3/2, l. (After Sándor, 1960).

sphere, is $\omega x^* \cos\theta$. If t is the time during which reflection of X-rays can occur during each passage of point P through the reflecting sphere, then

$$t \propto 1/\omega x^* \cos\theta \propto 1/\omega \sin 2\theta.$$

Now the blackening of the film due to the small volume round the point P is inversely proportional to t, i.e. directly proportional to $\sin 2\theta$. Thus the effect on the blackening of the film of the distortion in going from the reciprocal

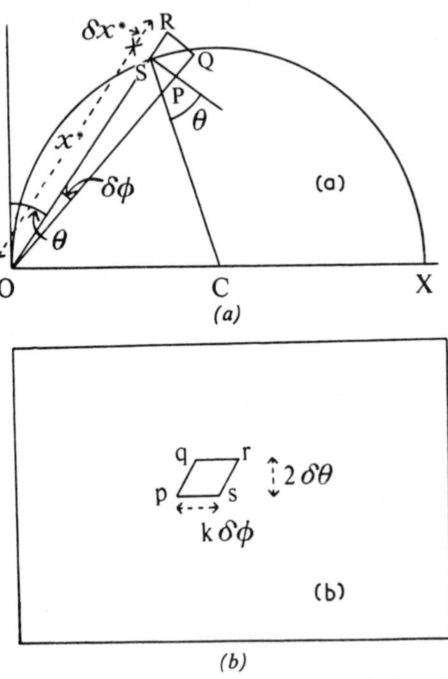

FIG. 5.10*a* Diagram showing the equivalent in reciprocal space of the parallelogram *pqrs* on the Weissenberg photograph shown in Fig. 5.10*b*.

FIG. 5.10*b* Diagram representing a Weissenberg photograph. The centre line of the photograph is supposed to coincide with the lower edge of the rectangle.

lattice to the photograph is just compensated by the Lorentz correction and both may be omitted. Since the geometry of equi-inclination photographs is the same as that for equatorial photographs, except for the change of diameter of the appropriate reflecting circle, the same arguments apply.

5:3.4.2 *Polarization correction*

When polarized X-rays are scattered from an electron the intensity of scattering depends on the angle between the electric vector of the X-rays and the plane containing the incident and scattered rays. For unpolarized X-rays, such as those normally used in the photographs discussed here, this angle varies between 0 and π. The ratio of the intensity of the scattered X-rays for which the angle is $\pi/2$ and 0 respectively is $\cos^2 2\theta$. The mean value for all directions of polarization is $(1+\cos^2 2\theta)/2$. The equi-inclination angle of a layer line is ν, and y (measured in degrees) is the coordinate of a point on the film measured along a line perpendicular to the trace of the direct beam. The stereogram, Fig. 5.11, shows the relation between various directions and angles. The incident beam travels in the direction represented by XX'; the reflected beam is represented by the point R lying on a great circle passing through XX'. The axis of oscillation of the crystal during the taking of the equi-inclination photograph is represented by the point A. The cone of rays transmitted by the circumferential opening in the screen is represented by the small circle XRB described about A. The angle between the reflected rays and the emergent beam $RX' = 2\theta$; the semi-angle of the cone of reflected rays $AR = \pi/2-\nu$, where ν is the angle through which the track of the camera is rotated from the normal-beam setting. For an equi-inclination photograph

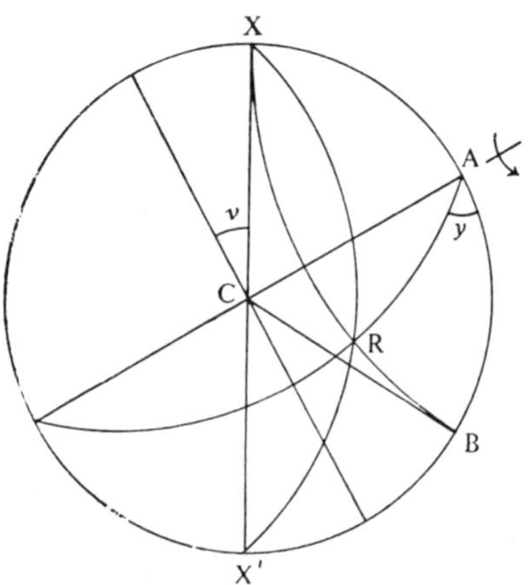

FIG. 5.11 Sreteogram showing the relation between various directions and angles involved in the equi-inclination Weissenberg photograph.

$$\sin \nu = \zeta/2,$$

where ζ is the height, in reciprocal space, of the layer line studied above the zero-level layer line. The angle $X'A = \pi/2+\nu$. The angle y measures the perpendicular distance of the point in the photograph from the central line. If the camera is of diameter 57·3 mm then this distance is $y/2$ mm.

From the spherical triangle $X'RA$ we have

$$\cos X'R = \cos RA.\cos X'A+\sin RA.\sin X'A.\cos RAX$$

or, inserting the symbols θ, ν, and y,

$$\cos 2\theta = -\sin^2 \nu+\cos^2 \nu.\cos y.$$

Thus to correct for the effect of polarization, using filtered radiation, the observed intensities must be multiplied by a factor

$$P = 1/\{1+(\cos^2 \nu\cdot\cos y-\sin^2 \nu)^2\}$$

When this polarization factor is plotted against the distance from the trace of the direct X-ray beam it gives a curve of the form shown in Fig. 5.12. The shape of this curve depends on v, as is shown by the curves A, B, C, D, E, which correspond respectively to values of v equal to 0°, 10°, 20°, 30°, 40°.

This correction can conveniently be applied on the Wooster microdensitometer. The output potential from that instrument, which operates a pen-recorder, is derived from a 1 kilo ohm potentiometer. Normally 25 volts is

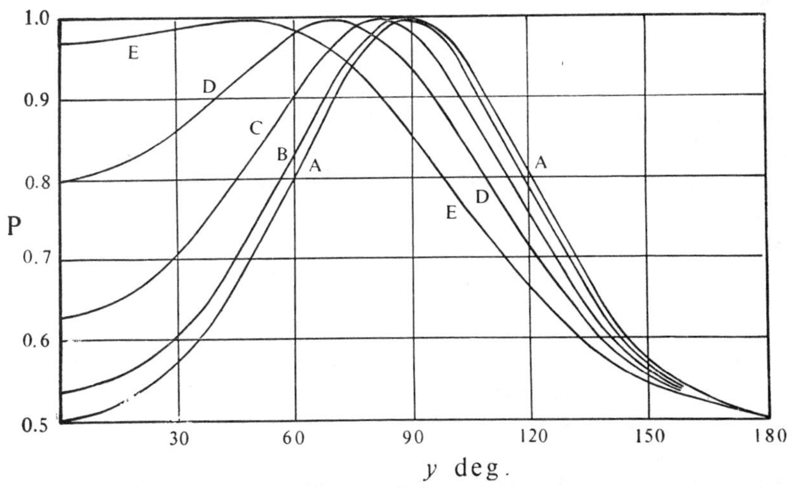

Fig. 5.12 Diagram showing, as ordinate, the factor by which the observed intensities must be multiplied on account of the polarization correction and, as abscissa, the distance from the trace of the direct beam on the Weissenberg photograph.

applied across this potentiometer but, for the purpose of applying the polarization correction, this voltage may be varied. If the voltage is made proportional to $1/\{1+(\cos^2 v.\cos y-\sin^2 v)^2\}$ for every traverse across the film parallel to the trace of the direct beam, then the contours are automatically corrected for the effects of polarization.

5:3.4.3 *Correction for Compton and collimator scattering*

The corrections mentioned above are capable of being precisely analyzed and applied with sufficient accuracy. This is often not true of the corrections now to be discussed. In addition to the diffuse scattering due to the thermal motion of atoms there is a diffuse scattering due to the Compton effect. James (1948, p. 461) gives an expression for the incoherent scattering of a single atom as

$$I_{\text{inc}} = \sum_k (1-|f_k|^2) - \sum_j^{j\neq k} \sum_k |f_{jk}|^2,$$

where f_k is the scattering factor associated with the k electron. The last term is generally a small correction term and the main effect is given by $(1-|f_k|^2)$. This quantity is zero at $\theta = 0°$ and rises continuously as θ increases. It can be calculated approximately though there is not agreement as yet as to the reliability of such calculations. The false assumption is that electrons in crystals may be considered as free electrons for the purpose of deriving the incoherent scattering.

The collimator scattering is purely instrumental and arises because of total and multiple reflections of the X-ray beam within the collimator. Even with the most careful attention to design of the collimator it is difficult to avoid a scattering which is especially troublesome at small angles of θ. The exposure times are generally long and consequently even small amounts of instrumental scattering become important. A photograph taken under identical conditions to that of the one being studied except that the crystal is omitted will enable the amount of collimator scattering to be determined. The polarization correction mentioned above does not apply either to the Compton or to the collimator correction. Strictly speaking, the Compton and collimator scattering should first be subtracted from the measured diffuse intensity and then the polarization correction applied. Because of the uncertainties in these corrections and because the Compton scattering increases with θ while the collimator scattering decreases with θ, Sándor treated these corrections as cancelling each other and no allowance was made for either. This procedure is somewhat arbitrary but after making corrections in this way the level of the background all over the contoured diagram (Fig. 5.13*a*) was found to be constant. This result is to be expected when the corrections are properly applied.

5:3.4.4 *Absorption correction*

Unless the crystal is in the form of a sphere the absorption of the scattered rays depends on the angle of deviation, 2θ. The absorption coefficient for anthracene is so low that with the size of crystal used in taking the Weissenberg photograph shown in Fig. 5.9, the maximum variation in the intensity of the diffusely reflected beams due to this cause was 10 per cent. The correction for this was neglected in view of other larger sources of error, but it could be applied if the shape and size of the cross section of the crystal were accurately measured. For diffuse reflections the usual difficulty in determining the actual linear absorption coefficient due to the effect of extinction does not arise, since there is no coherence in the scattered beams.

5:3.5 *Comparison for anthracene of experimental and theoretical contoured diagrams*

The diffuse regions of interlayer Weissenberg photographs were studied using the microdensitometer as described above and from the two halves of the photograph for the section of reciprocal space containing point h, 3/2, l the two contoured diagrams of Fig. 5.13*a* were obtained. An important feature is the splitting of the 'hills' having maximum heights of 9 and 12 into a pair of parallel ridges. This effect can be clearly seen in Fig. 5.9 on the diffuse streaks near the centre line of the photograph.

The DFT for the h, 3/2, l section was calculated for one molecule in the unit cell by means of isotropic thermal vibrations. For these computations an electronic digital computer was essential. To obtain the effect of both molecules in the unit cell the contoured diagram so obtained was rotated through 180° about its centre and the summation of the heights at every point common to the two maps made. The result is shown in Fig. 5.13*b*. A similar contoured diagram which took into account anisotropic translational and rotational vibrations of the molecules was drawn, but this did not differ significantly from Fig. 5.13*b*. Comparing Figs. 5.13*a* and 5.13*b* it will be seen that the peaks occur in the same positions and are very nearly of the same height. Thus in the right-hand half of Fig. 5.13*a* we have peak-heights of 14, 9, 12, 11 and the corresponding values in Fig. 5.13*b* are 14, 6, 11, 10 respectively. Thus we may conclude that the assumption of independent molecular vibrations gives an adequate account of the observed diffuse scattering. The salient feature left unexplained is the splitting of the peaks of heights 9 and 12 in Fig. 5.13*a*.

5:3.6 *Orientation of the molecules*

The peaks of Figs. 5.13*a* and 5.13*b* correspond to the sections by the plane (0, 3/2, 0) of the Fourier transforms of the two molecules in the unit cell. These transforms consist of diffuse rods arranged hexagonally with their lengths perpendicular to the planes of the molecules. The normal to the plane (010) is inclined at an angle of 61° to the length of these diffuse rods and this accounts for the elliptical form of the peaks in Figs. 5.13*a* and 5.13*b*. A further consequence is that the projection of these diffuse rods on the plane (010) is parallel to the major axes of the diffuse peaks. A difficulty arises here because there are two sets of superposed peaks, one due to each of the two molecules in each unit cell. This causes some scatter in the direction of the major axes of the diffuse peaks. However, it is possible to determine the direction of the projection of the normal to the plane of the molecule on the plane (010) with an accuracy of about $\pm 2°$. The line marked M in Figs. 5.13*a* and 5.13*b* was obtained from the crystal structure determinations and it will be seen to correspond closely with the long axes of the diffuse peaks. Similar sections of reciprocal space parallel to other planes than (010) enable other projections of the diffuse rods to be studied, and in such a case as anthracene the diffuse reflections afford a valuable means of orientating the planes of the molecules within the unit cell.

5:4 **Conclusion**

In this chapter an account has been given of the main outlines of the theoretical and experimental work done on certain selected molecular crystals.

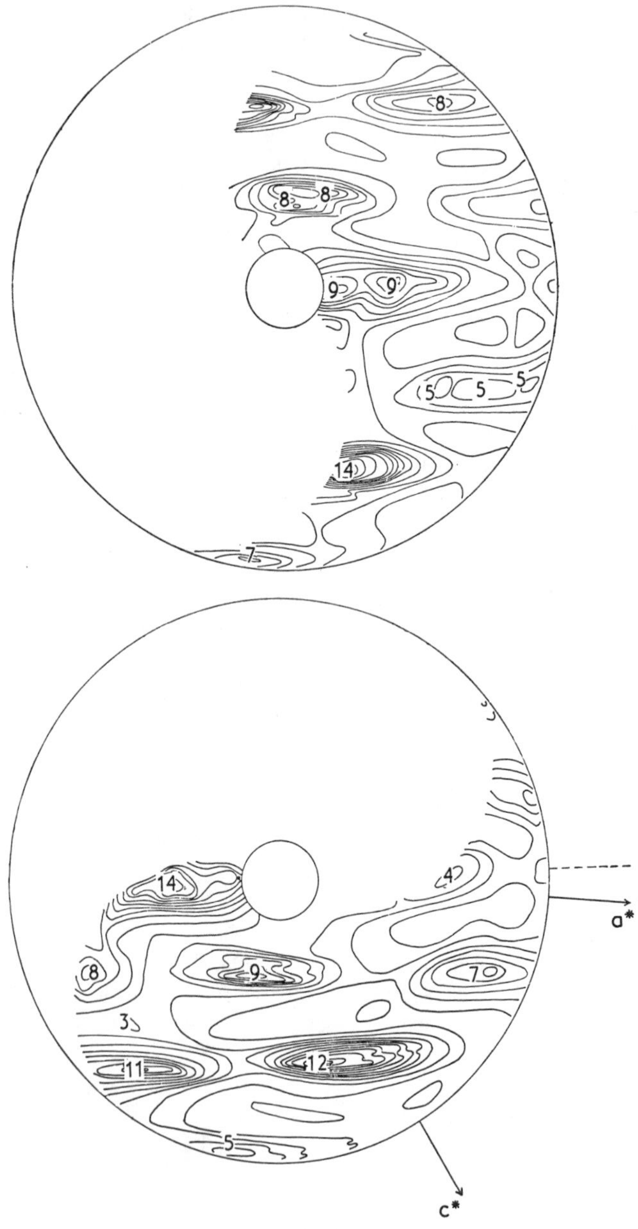

FIG. 5.13*a* Experimentally determined distribution in reciprocal space for the section *h*, 3/2, *l*, obtained from Fig. 5.9. The dotted line marked *M* corresponds to the projection on (010) of the normal to the plane of the molecule.

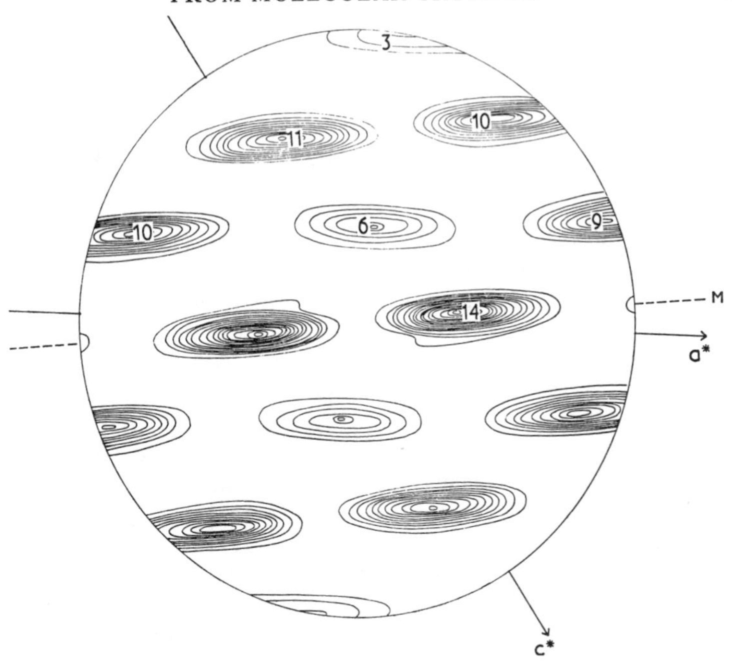

FIG. 5.13*b* Calculated distribution corresponding to Fig. 5.10*a*. (After Sándor, 1960).

The reason for this selection is that a correlation between observation and theory is only possible at present for molecules that have simple well-defined shapes and relatively rigid frameworks. Notable omissions from the crystals discussed are ice and urea. Both of these crystals give striking diffuse patterns which cannot at present be interpreted. One probable reason for this is that the internal vibrations play a much larger part relative to the external vibrations than is the case, for instance, in anthracene. Another difficulty with such small molecules is that the pattern of the diffuse scattering (Owston, 1949) does not correspond with any Fourier transform or *K*-surface. Diffuse rods and plates frequently join neighbouring relps and no theoretical explanation of this is available. Some of these effects may be due to types of disorder of the kind discussed in Chapter IV but it is doubtful whether many of the unexplained effects can be accounted for in this way.

CHARTS FOR SETTING CRYSTALS ON A DIFFRACTOMETER; i-, ϕ-CHARTS

In using a diffractometer for the study of diffuse X-ray reflections, it is necessary to determine which point, or element of volume, in reciprocal space is contributing to the measured effect. The orientation of the crystal and the aperture in front of the detector completely determine this. A normal Bragg reflection from atomic planes which are parallel to the surface of the crystal plate is characterized by the settings i and ϕ for the crystal and detector, respectively, where $i = \theta$ and $\phi = 2\theta$. On the charts below the line for which these equations are true is a straight horizontal line approximately mid-way down the page. This line contains the white radiation streaks passing through relps. The traces of powder lines which often occur lie along the constant-ϕ curves which are nearly vertical. In printing these curves the original scale may be changed. It is always possible to find the true radius, R, of the reflecting sphere for a given i-, ϕ-chart by noting that the difference in the radii of two ϕ-circles, having values ϕ_1 and ϕ_2 respectively, is $2R(\sin \phi_2/2 - \sin \phi_1/2)$ (See also Fig. 1.29 for the ϕ-range 20°–47°.)

Charts for determining the skew (absorption) correction

When the angle of incidence on a face of a crystal is not equal to the angle of reflection, the measured intensity of reflection must be corrected for the

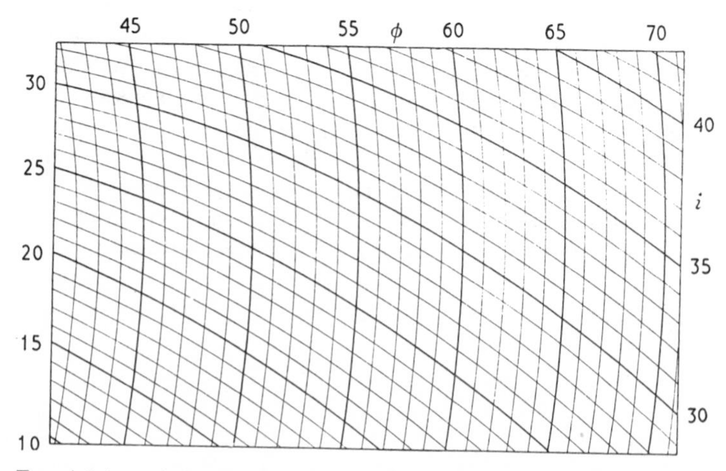

FIG. A.1.1 i-, ϕ-chart for range in ϕ of 42°–70°. (After Ramachandran, 1949).

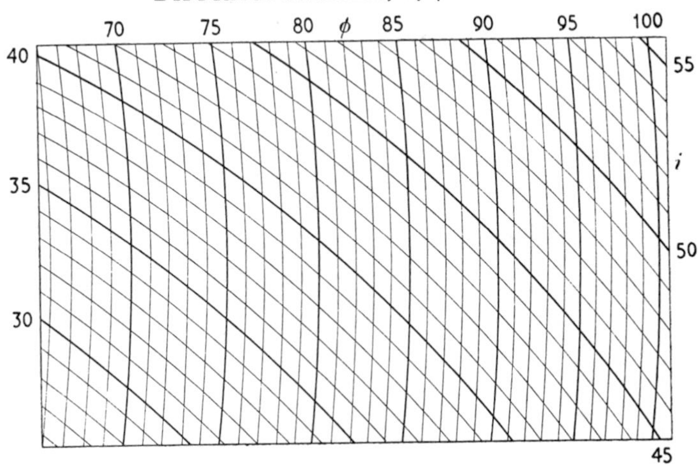

FIG. A.1.2 *i*-, ϕ-chart for range in ϕ of 66°–100°.

absorption in the crystal. The following charts are constructed on the same scale as the *i*-, ϕ-charts and may be superposed on the same drawing of the reciprocal points as those charts. The ranges cover the ϕ-values 20°–47° (see Fig. 1.32), 41°–71°, 66°–100°, and give the factor by which the observed intensity must be multiplied to get the quantity which would be observed if the reflecting planes were parallel to the actual surface of the crystal.

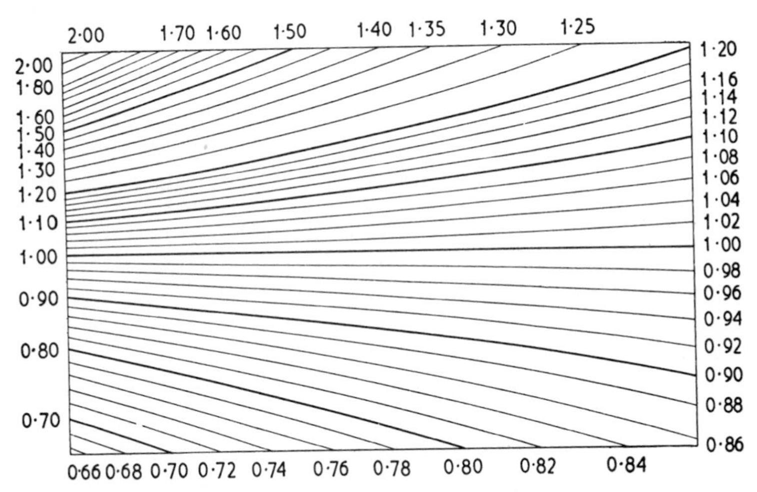

FIG. A.1.3 Chart of absorption correction. Region 2. 41°–71°. (After Ramachandran and Wooster, 1951*a*).

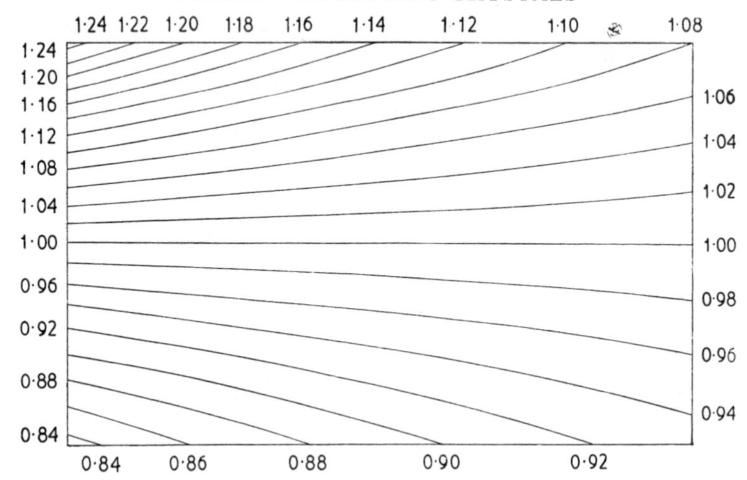

FIG. A.1.4. Chart of absorption correction. Region 3. 66°–100°.

APPENDIX II

A METHOD OF COMPUTING $\bar{\rho}$-, $\bar{\phi}$-, AND K*-CHARTS

A method of deriving the charts similar to those of Figs. 1.23 and 1.24 was given by Hoerni and Wooster (1952*c*) and the following method is based on their work. To illustrate the process the x, y coordinates of one line in Fig. 1.23, namely, that for $\bar{\rho} = 40°$, will be evaluated. The distance of the relp from the reflecting circle, *PN* in Figs. 1.21 and 1.22, is put equal to 2 cm, and the radius of the reflecting sphere, *NI*, to 100 cm, so that $s = +0{\cdot}02$. In Fig. A.2.1 is shown a section through the reflecting sphere containing the line *PI* and the point *Q* (the letters *P*, *I*, *Q* have the same significance as in Figs. 1.21 and 1.22). The angles *QPI* and *QIP* are denoted $(\pi/2-e)$ and f respectively. The inclination of the plane *QPI* to the plane *PZI*, Fig. 1.21, which is the vertical plane through *PI*, is denoted $\bar{\psi}$. The length *NQ* and the angle $\bar{\psi}$ are polar coordinates of a point on the chart from which the cartesian coordinates x, y can readily be obtained. It is important to notice that because the angle f is seldom larger than a few degrees, the line *NQ* may be assumed to be of the same length as the arc *NQ* or the line *NF*, which is obtained by producing *IQ* to cut the tangent plane to the reflecting sphere at *N*. In the equatorial plane the radius of the circle, corresponding to a given combination of *PQ* and e, is of length equal to the arc *NQ*. In the vertical plane the corresponding radius is *NF*. We shall not attempt to deal with the slightly oval-shaped curves which replace the circles when the angle f becomes larger than a few degrees.

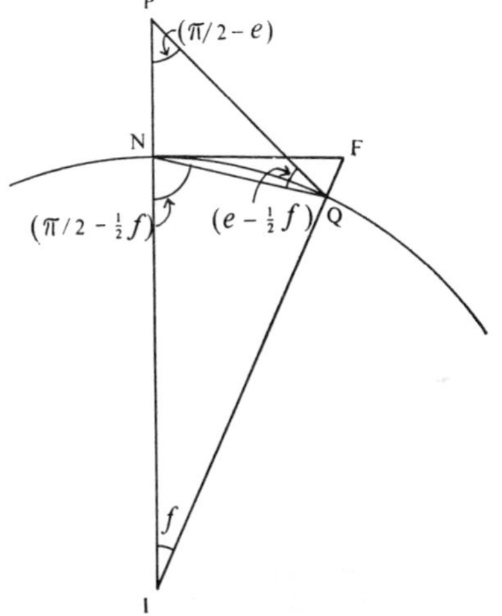

FIG. A.2.1 Diagram showing the relations between angles and lengths required in the computation of $\bar{\rho}$-, $\bar{\phi}$-charts.

The accuracy over the whole of the chart in Fig. 1.23 is high enough for graphical work. For chart with $s = +0{\cdot}05$ the K-circles are only circular towards the middle of the chart. On the periphery the curves become ovaloid, though even in this case the departure from true circularity is only about 2 mm at a distance of 10 cm from the centre ($NP = 2$ cm).

$\bar{\rho}$-, $\bar{\Phi}$-chart

Fom Fig. A.2.1 we see that angle $NQP = (e-f/2)$ and hence, neglecting second powers of f, we obtain

$$PN/NQ = \sin(e-f/2)/\cos e$$
$$= (\tan e)-f/2.$$

The length of PN may be taken of any value to suit the scale of the chart, e.g. 2 cm, 5 cm, or 10 cm, but here we shall put $PN = 2$ cm.

Further we have,

$$f = NQ/NI$$

and, since $NI = 100$ cm,

$$f = NQ/100.$$

Thus, the final equation for deriving NQ is

$$2/NQ = (\tan e)-NQ/200. \tag{A.2.1}$$

In Fig. A.2.2 is shown a stereogram giving the angular relationships between the lines drawn in Fig. 1.22. The point Q represents the direction PQ in Fig. A.2.1. From the Napierian triangles it may be shown that

$$\sin e = \sin \bar{\Phi} . \sin \bar{\rho}, \tag{A.2.2}$$

$$\tan \bar{\psi} = \tan \bar{\rho} . \cos \bar{\Phi}. \tag{A.2.3}$$

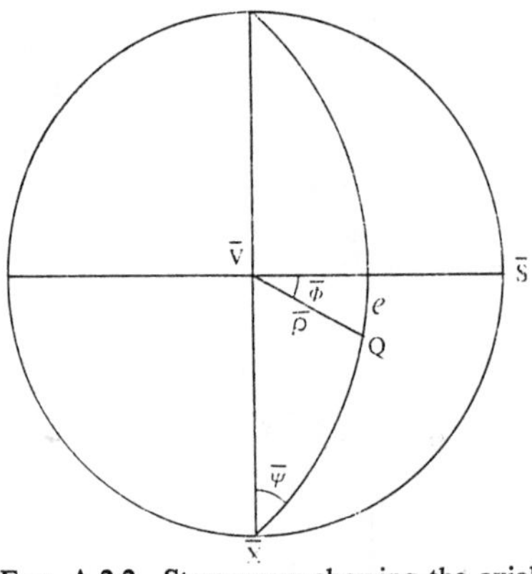

FIG. A.2.2 Stereogram showing the axial reference system $\bar{X}$, $\bar{V}$, $\bar{S}$ and the angles $\bar{\rho}$, $\bar{\Phi}$, e, and $\bar{\psi}$, defining the direction Q corresponding to PQ in Fig. A.2.1.

To calculate one line of the chart we put $\bar{\rho} = 40°$ and give $\bar{\Phi}$ the values 90°, 70°, 50°, 30° in turn. From equation (A.2.2) we obtain e for these combinations and insert them in equation (A.2.1). This equation need not be treated as a quadratic in NQ but the value of NQ may be found for any value of e by a simple process of successive approximations. From equation (A.2.3) the angle $\bar{\psi}$ can be calculated for the given combinations of $\bar{\rho}$ and $\bar{\Phi}$. Finally, we obtain the cartesian coordinates from the equations

$$\left.\begin{aligned} x &= NQ . \sin \bar{\psi} \\ y &= NQ . \cos \bar{\psi} \end{aligned}\right\}. \tag{A.2.4}$$

The numerical evaluations which follow illustrate the method described above.

$\bar{\rho} = 40°$

	90°	70°	50°	30°
e	40° 0′	37° 10′	29° 30′	18° 45′
$\bar{\psi}$	0° 0′	16° 1′	28° 21′	36° 0′
f	0·0242	0·0268	0·0365	0·0651 rad.
NQ	2·42	2·68	3·63	6·51 cm
x	0·00	0·74	1·73	3·83 cm
y	2·42	2·58	3·21	5·27 cm

K-chart*

To construct a K^*-chart such as that shown in Fig. 1.24 we proceed as follows. Particular values are assigned to K^* and we calculate the corresponding values of the angles e and f. From these the radius NQ can be found for each of the chosen values of K^*.

From triangle PNQ, Fig. A.2.1, we have

$$\sin(e-f/2) = \frac{NP}{PQ}.\cos f/2$$

and, neglecting squares of f, we have

$$\sin e = \frac{NP}{K^*} + (\cos e).f/2. \qquad \text{(A.2.5)}$$

Further, from triangle PQI,

$$\frac{PQ}{\sin f} = \frac{IQ}{\cos e}$$

hence,

$$f = \frac{K^*}{100}.\cos e.$$

Substituting this value in equation (A.2.5) we obtain

$$\sin e = \frac{NP}{K^*} + \frac{K^*}{200}.\cos^2 e. \qquad \text{(A.2.6)}$$

By successive approximations the angle e can be found for each value of K^*. Then from the equation (A.2.1) we can find NQ, which is the radius of the circle corresponding to the given value of K^*.

The following numerical example is given to illustrate the process.

K	2·0	3·0	4·0	5·0 cm
e	90° 0′	42° 27′	30° 59′	24° 52′
NQ	0·00	2·22	3·43	4·54 cm

(Note. The numbers in Fig. 1.24 are five times greater than the corresponding values calculated here, owing to the distance NP being made five times greater.)

If s is negative the angle NQP in Fig. A.2.1 becomes $(e+f/2)$ and in equations (A.2.1) and (A.2.6) the small second terms change sign.

APPENDIX III

CONVENTION ON THE SUBSCRIPTS FOR STRESS AND STRAIN TENSORS

DIFFERENT authors have conflicting conventions regarding the definition of the components of stress. The older description involved two letters, e.g. X_y in which the capital X referred to the direction of the stress component and the subscript (small) letter referred to the axis normal to the face on which the force component acts. This convention was followed by Kirchhoff (1876) p. 390, Liebisch (1891) p. 547, Love (1892) p. 79, Voigt (1910) p. 164, and Cady (1946) p. 47. Also using capital and small letters, Seitz (1943) p. 64 used the reversed convention, i.e. the capital letter referred to the face on which the force acts and the small letter to the direction of the force component.

The use of a tensor notation has given rise to the replacement of the letters by numbers so that stress components are often denoted τ_{ik}, where i and k may take any of the values 1, 2, or 3. Some authors have made the first suffix to correspond with the directions of the force component and the second with the axis normal to the face in which it acts. This group includes Frocht (1941) p. 4, Mason (1950) p. 21, and Nye (1957) p. 82. The opposite convention has been adopted by Jeffreys (1931) p. 74, Wooster (1938) p. 233, Sokolnikoff (1946) p. 37, de Jong (1959) p. 196. In view of this unfortunate lack of agreement, the author has thought it best to adhere to the same convention used in *A Text-book on Crystal Physics* (1938).

BIBLIOGRAPHY AND AUTHOR INDEX

de Acha, A. *see* Amorós, J. L. *et al.* (1958). 158, 159, 164

Ahmed, M. S. (1952) *Acta cryst.* **5**, 587. 158

Allison, S. K. *see* Compton, A. H. (1935). 53, 97

Amorós, J. L. and Canut, M. L. (1957) *Acta cryst.* **10**, 794, 160

— and — (1958*a*) *Bol. R. Soc. Esp. Hist. Nat.* (G) **56**, 305. 154

— and — (1958*b*) *P. Dep. Crist. Min.* (*Madrid*) **4**, 7. 160, 161, 162

—, —, Annaka, S., and de Acha, A. (1958) Technical (Final) Report. Contract AF 61 (514) - 1146. 158, 159, 164

—, —, and Bujosa, A. (1957) First Annual Report. Contract AF 61 (514) - 1146. 154

— *see* Annaka, S. (1960). 165

— *see* Canut, M. L. (1960). 94

Annaka, S. (1956) *J. Phys. Soc. Japan* **11**, 937. 59

— and Amorós, J. L. (1960) *J. Phys. Soc. Japan* **15**, 356. 165

— *see* Amorós, J. L. *et al.* (1958). 158, 159, 164

Averbach, B. L. *see* Flinn, P. A. (1951). 2, 116

— *see* — *et al.* (1954). 116

— *see* Herbstein, F. H. *et al.* (1956). 116, 146

— *see* Houska, C. R. (1959). 117

— *see* Warren, B. E. *et al.* (1951). 146, 150, 152

Bagaryatskii, Yu. A. (1951) *Dokl. Akad. Nauk S.S.S.R.* **77**, 45. 114, 115

— (1952) *Dokl. Akad Nauk S.S.S.R.* **87**, 559. 115

Bagchi, S. N. *see* Hoseman, R. (1954). 113

Batterman, B. W. (1957) *J. appl. Phys.* **28**, 556. 116, 145, 146, 151

Baumgärtner, F. *see* Hoppe, W. (1957). 160

Begbie, G. H. (1947) *Proc. Roy. Soc.* A **188**, 189. 2

— and Born, M. (1947) *Proc. Roy. Soc.* A **188**, 179. 2

Bergmann, L. (1938) *Ultrasonics.* London: G. Bell & Sons. 57

— *see* Schaefer, Cl. (1934; 1935). 57

Bijvoet, J. M. and Nieuwenkamp, W. (1933) *Z. Krist.* **86**, 466. 113

Borie, B. (1957) *Acta cryst.* **10**, 89. 117

— and Warren, B. E. (1956) *J. appl. Phys.* **27**, 1562. 116

— Jr. *see* Herbstein, F. H. *et al.* (1956). 116, 146

Born, M. (1942*a*) *Proc. Phys. Soc.* **54**, 362. 2, 59

— (1942*b*) *Proc. Roy. Soc.* A **180**, 397. 2, 59

— (1943) *Rep. Prog. Phys.* **9**, 294. 2

— and Karman, Th. von (1912) *Physik. Z.* **13**, 297. 17

— and — (1913) *Physik. Z.* **14**, 15, 65. 17

— and Sarginson, K. (1941) *Proc. Roy. Soc.* A **179**, 69. 2

— *see* Begbie, G. H. (1947). 2

Bown, M. G. and Gay, P. (1958) *Z. Krist.* **111**, 1. 114

Bragg, W. L. and Lipson, H. (1943) *J. sci. Instrum.* **20**, 110. 113

BRAGG, W. L. and WILLIAMS, E. J. (1935) *Proc. Roy. Soc.* A **151**, 540. 145
BRIDGMAN, G. W. (1938) *Proc. Amer. Acad. Arts Sci.* **72**, 207. 85
— (1945) *Proc. Amer. Acad. Arts Sci.* **76**, 9. 85, 88
BRINDLEY, G. W. OUGHTON, B. M., and ROBINSON, K. (1950) *Acta cryst.* **3**, 408. 113
— and RIDLEY, P. (1938) *Proc. Phys. Soc.* **50**, 757. 1
BUJOSA, A. *see* AMORÓS, J. L. *et al.* (1957). 154
BURGERS, W. G. and HIOK, T. K. (1946) *Physica* **11**, 353. 3

CADY, W. G. (1946) *Piezoelectricity*. New York: McGraw-Hill. 181
CANUT, M. L. and AMORÓS, J. L. (1960) *Bol. R. Soc. Esp. Hist. Nat* (G) **57**, 43. 94
— *see* AMORÓS (1957). 160
— *see* — *et al.* (1957). 154
— *see* — (1958*a*). 154
— *see* — (1958*b*). 160, 161, 162
— *see* — *et al.* (1958). 158, 159, 164
CARTZ, L. (1955*a*) *Proc. Phys. Soc.* B **68**, 951. 94
— (1955*b*) *Proc. Phys. Soc.* B **68**, 957. 94
CATICHA-ELLIS, S. and COCHRAN, W. (1957) *Acta cryst.* **10**, 826. 114, 132
CHARLESBY, A., FINCH, G. I.,and WILMAN, H. (1939) *Proc. Phys. Soc.* **51**, 479. 164
CHAYES, F. (1956) *Carnegie Institute of Washington Year Book* **56**, 151. 114
CHIKAWA, J. *see* TANAKA, K. *et al.* (1959). 54
CHIPMAN, D. and WARREN, D. E. (1950) *J. appl. Phys.* **21**, 696. 54, 116
CHIPMAN, D. R. (1956) *J. appl. Phys.* **27**, 739. 116, 146
— and PASKIN, A. (1959*a*) *J. appl. Phys.* **30**, 1992. 117
— and — (1959*b*) *J. appl. Phys.* **30**, 1998. 117
CHRISTOFFEL, E. B. (1877) *Ann. di. Mat.* (Series 2) **8**. 63
— (1910) *Ges. math. Abhandl.* **2**, 81. 63
COCHRAN, W. (1956) *Acta cryst.* **9**, 259. 116, 146, 154
— and KARTHA, G. (1956*a*) *Acta cryst.* **9**, 941. 116
— and — (1956*b*) *Acta cryst.* **9**, 944. 116, 151
— *see* CATICHA-ELLIS, S. (1957). 114, 132
COLE, H. (1953) *J. appl. Phys.* **24**, 482. 96
— and WARREN, B. E. (1952) *J. appl. Phys.* **23**, 335. 2, 96, 112
COMPTON, A. H. and ALLISON, S. K. (1935) *X-rays in Theory and Experiment*. New York: Van Nostrand. 53, 97
CORRE, Y. LE (1953) *Bull. Soc. franç. Minéral. Crist.* **76**, 464. 56
COWLEY, J. M. (1950*a*) *J. appl. Phys.* **21**, 24. 115, 146
— (1950*b*) *Phys. Rev.* **77**, 669. 115
CRUICKSHANK, D. W. J. (1956) *Acta cryst.* **9**, 915. 156
— (1957) *Acta cryst.* **10**, 470. 156
CURIEN, H. (1952*a*) *Bull. Soc. franç. Minéral. Crist.* **75**, 197. 2, 59, 96, 99, 110
— (1952*b*) *Acta cryst.* **5**, 393. 2, 59, 96, 110
— (1952*c*) *Bull. Soc. franç. Mineral. Crist.* **75**, 343. 2, 96, 110
— (1952*d*) Thesis. Paris. 2, 96, 99, 110, 111
— (1958) *Rev. mod. Phys.* **30**, 232. 97

DEAS, H. D. (1952) *Acta cryst.* **5**, 542. 3, 117
DEBYE, P. (1912) *Ann. Phys., Lpz.* **39**, 789. 95
— (1913*a*) *Verh. dtsch. phys. Ges.* **15**, 678. 1

DEBYE, P. (1913*b*) *Verh. dtsch. phys. Ges.* **15**. 738 1
— (1913*c*) *Verh. dtsch. phys. Ges.* **15**, 857. 1
— (1914) *Ann. Phys., Lpz.* **43**, 49. 1
DOI, K. (1957) *Bull. Soc. franç. Minéral. Crist.* **80**, 325. 115
— (1960) *Acta cryst.* **13**, 45. 115, 140, 142
DORNBERGER-SCHIFF, K. (1956) *Acta cryst.* **9**, 593. 114, 119, 123
— (1957) *Acta cryst.* **10**, 271. 114

EDMUNDS, I. G., HINDE, R. M., and LIPSON, H. (1947) *Nature* **160**, 304. 115, 138
— *see* STEEPLE, H. (1956). 116
EDWARDS, O. S. and LIPSON, H. (1942) *Proc. Roy. Soc.* A **180**, 268. 113
EKSTEIN, H. (1945) *Phys. Rev.* **68**, 120. 3
ELLIOTT, R. J. (1960) Clarendon Laboratory, Oxford. Ref. No. 23/60. 114, 131
EMDE, F. *see* JAHNKE, E. (1945). 101

FAXEN, H. (1923) *Z. Phys.* **17**, 266. 1
FINCH, G. I. *see* CHARLESBY, A. *et al.* (1939). 164
FLINN, P. A. and AVERBACH, B. L. (1951) *Phys. Rev.* **83**, 1070. 2, 116
—, —, and RUDMAN, P. S. (1954) *Acta cryst.* **7**, 153. 116
FRANK, F. C. (1956) *Proc. Roy. Soc.* A **237**, 168. 114
FROCHT, M. M. (1941) *Photoelasticity*, Vol. 1. New York: J. Wiley & Sons. 181
FRUHLING, A. (1951) *Ann. Phys., Paris* **6**, 401. 157

GAY, P. *see* BOWN, M. G. (1958). 114
GEISLER, A. H. *see* NEWKIRK, J. B. *et al.* 114
GEROLD, V. (1954) *Z. Metallk.* **45**, 599. 140, 142
— (1958) *Acta cryst.* **11**, 230. 140, 142
GEVERS, R. (1954*a*) *C.R. Acad. Sci., Paris* **238**, 1827. 113
— (1954*b*) *Acta cryst.* **7**, 337. 113
GOENS, E. (1933) *Ann. Phys., Lpz.* **17**, 233. 107
GUINIER, A. (1938) *C.R. Acad. Sci., Paris* **206**, 1641. 114, 140
— (1942) *J. Phys. Radium* **3**, 124. 114, 140
— (1943*a*) *Métaux et Corrosion* **18**, 209. 114
— (1943*b*) *J. Chim. phys.* **40**, 133. 114
— (1944) *Rev. Métall.* 1. 114
— (1945*a*) Radiocristallographie. Dunod. Paris. 3
— (1945*b*) *Proc. Phys. Soc.* **57**, 310. 3, 115, 141, 142, 146
— (1949) *Physica* **15**, 148. 140
— (1950) *C.R. Acad. Sci., Paris* **231**, 655. 114, 140
— (1952) *Acta cryst.* **5**, 121. 115, 140

HALL, J. H. (1942) *Phys. Rev.* **61**, 158. 1
HARGREAVES, M. E. (1951) *Acta cryst.* **4**, 301. 114
HARKER, D. *see* KASPER, J. S. *et al.* (1950). 117
HAUSSÜHL, S. (1958) *Acta cryst.* **11**, 58. 158
HENDRICKS, S. B. (1938) *Z. Krystallogr.* **99**, 264. 113
— (1939) *Amer. Min.* **24**, 529. 113
— (1940) *Phys. Rev.* **57**, 448. 113
— and JEFFERSON, M. E. (1938) *Amer. Min.* **23**, 851. 113

HENDRICKS, S. B. and JEFFERSON, M. E. (1939) *Amer. Min.* **24**, 729. 113
— and TELLER, E. (1942) *J. chem. Phys.* **10**, 147. 113
HERBSTEIN, F. H., BORIE, B. S. JR., and AVERBACH, B. L. (1956) *Acta cryst.* **9**, 466. 116, 146
HINDE, R. M. *see* EDMUNDS, I. G. *et al.* (1947). 115, 138
— *see* TAYLOR, C. A. *et al.* (1951). 115, 138
HIOK, T. K. *see* BURGERS, W. G. (1946). 3
HIRABAYASHI, M. and OGAWA, S. (1956) *J. Phys. Soc. Japan* **11**, 907. 116
HOERNI, J. A. (1952) Thesis. Cambridge. 36
— and WOOSTER, W. A. (1952*a*) *Experientia* **8**, 297. 114
— and — (1952*b*) *Acta cryst.* **5**, 386. 25, 59
— and — (1952*c*) *Acta cryst.* **5**, 626. 31, 177
— and — (1953) *Acta cryst.* **6**, 543. 37
— and — (1955) *Acta cryst.* **8**, 187. 114, 127, 132
HOPPE, W. (1956*a*) *Z. Krist.* **107**, 406. 157
— (1956*b*) *Z. Krist.* **107**, 433. 163
— (1957) *Z. Krist.* **108**, 335. 163
— and BAUMGÄRTNER, F. (1957) *Z. Krist.* **108**, 328. 160
— and LENNÉ, H. N., and MORANDI, G. (1957) *Z. Krist.* **108**, 321. 162, 163
HOSEMANN, R. and BAGCHI, S. N. (1954) *Phys. Rev.* **94**, 71. 113
HOSHINO, S. (1957) *J. Phys. Soc. Japan* **12**, 315. 114
HOUSKA, C. R. and AVERBACH, B. L. (1959) *J. appl. Phys.* **30**, 1532. 117
HUANG, K. (1947) *Proc. Roy. Soc.* A **190**, 102. 116
HUSE, G. *see* POWELL, H. M. (1943). 117

JACOBSEN, E. H. (1955) *Phys. Rev.* **97**, 654. 2, 59
JAGODZINSKI, H. (1949*a*) *Acta cryst.* **2**, 201. 113
— (1949*b*) *Acta cryst.* **2**, 208. 113
— (1949*c*) *Acta cryst.* **2**, 298. 113
JAHN, H. A. (1942*a*) *Proc. Roy. Soc.* A **179**, 320. 70
— (1942*b*) *Proc. Roy. Soc.* A **180**, 476. 69
JAHNKE, E. and EMDE, F. (1945) *Tables of Functions with Formulae and Curves*. New York: Dover. 101
JAMES, R. W. (1948) *The Optical Principles of the Diffraction of X-rays*. London: G. Bell & Sons. 16, 53, 59, 68, 94, 153, 156, 169
— and SAUNDER, D. H. (1947) *Proc. Roy. Soc.* A **190**, 518. 117
— and — (1948) *Acta cryst.* **1**, 81. 117
JEFFERSON, M. E. *see* HENDRICKS, S. B. (1938), (1939). 113
JEFFREY, J. W. (1953) *Acta cryst.* **6**, 821. 114, 118
JEFFREYS, H. (1931) *Cartesian Tensors*. Cambridge University Press. 181
— and JEFFREYS, B.S. (1946) *Methods of Mathematical Physics*. Cambridge University Press. 100
JOEL, N. and WOOSTER, W. A. (1960) *Acta cryst.* **13**, 516. 56
JONES, F. W. and SYKES, C. (1938) *Proc. Roy. Soc.* A **166**, 376. 115
— *see* SYKES, C. (1936). 115
DE JONG, W. F. (1959) *General Crystallography*. San Francisco: W. H. Freeman. 181
JOST, K. H. *see* KRUMBIEGEL, J. (1955). 114
JOYNSON, R. E. (1954) *Phys. Rev.* **94**, 851. 2

KAKINOKI, J. and KOMURA, Y. (1952) *J. Phys. Soc. Japan* **7**, 30. 3, 116
— and — (1954*a*) *J. Phys. Soc. Japan* **9**, 169. 116
— and — (1954*b*) *J. Phys. Soc. Japan* **9**, 177. 116

KARMAN, TH. v. *see* BORN, M. (1912, 1913). 17
KARTHA, G. *see* COCHRAN, W. (1956*a*). 116
— *see* — (1956*b*). 116, 151
KASPER, J. S., LUCHT, C. M., and HARKER, D. (1950) *Acta cryst.* **3**, 436. 117
KATAYAMA, K. *see* TANAKA, K. *et al.* (1959). 54
KETELAAR, J. A. (1934) *Z. Kryst.* **88**, 26. 113
KIRCHHOFF, G. (1876) *Vorlesungen über math. Physik. Mechanik.* Leipzig: Teubner. 181
KOLONTSOVA, E. V. (1950) *Dokl. Akad. Nauk S.S.S.R* **75**, 189. 113
KOMURA, Y. *see* KAKINOKI, J. (1952) 3, 116
— *see* — (1954*a*, *b*). 116
KRISHNAN, R. S. and RAMACHANDRAN, G. N. (1945) *Nature* **155**, 234. 2
KRIVOGLAZ, M. A. (1957) *Zh. eksp. teor. Fiz.* **32**, 1368. 3
— (1958) *Zh. eksp. teor. Fiz.* **34**, 204. 3
KRUMBIEGEL, J. and JOST, K. H. (1955) *Z. Naturforsch.* **10*a***, 526. 114

LANDAU, L. (1936) *Phys. Z. Sowjetunion* **12**, 579. 116
LAUE, M. VON (1943) *Röntgenstrahl Interferenzen.* Ann Arbor, Michigan: Edwards Brothers. 146
LAVAL, J. (1938) *C.R. Acad. Sci., Paris* **207**, 169. 58
— (1939) *C.R. Acad. Sci., Paris* **208**, 1512. 1, 58, 59
— (1951) *C.R. Acad. Sci., Paris* **232**, 1947. 56
— (1954*a*) *J. Phys. Radium* **15**, 545. 59
— (1954*b*) *J. Phys. Radium* **15**, 657. 59
— (1959) *J. Phys. Radium* **20**, 1. 59
— *see* MAUGUIN, C. (1939). 58
LENNÉ, H. N. *see* HOPPE, W. *et al.* (1957). 162, 163
LIEBISCH, TH. (1891) *Physikalische Krystallographie*, Leipzig: Veit & Co. 181
LIFSHITS, I. M. (1948) *J. exp. theor. Phys. U.S.S.R.* **18**, 293. 59
LIPSON, H. and TAYLOR, C. A. (1951) *Acta cryst.* **4**, 458. 155
— and — (1958) *Fourier Transforms and X-ray Diffraction.* London: G. Bell & Sons. 4
— *see* BRAGG, W. L. (1943). 113
— *see* EDMUNDS, I. G. *et al.* (1947). 115, 138
— *see* EDWARDS, O. S. (1942). 113
— *see* TAYLOR, C. A. *et al.* (1951). 115, 138
LONGUET-ESCARD, J. and MÉRING, M. J. (1957) *Acta cryst.* **10**, 819. 114
LONSDALE, K. (1942*a*) *Proc. Roy. Soc.* A **179**, 315. 2
— (1942*b*) *Proc. Phys. Soc.* **54**, 314. 59, 160
— (1942*c*) *Nature* **149**, 698. 2, 59
— (1943) *Rep. Prog. Phys.* **9**, 256. 2
— (1945) *Nature* **155**, 572. 2, 3
— (1948) *Acta cryst.* **1**, 12. 59
— and SMITH, H. (1941) *Proc. Roy. Soc.* A **179**, 8. 59
LOVE, A. E. H. (1892) *The Mathematical Theory of Elasticity.* Cambridge University Press. 181
LUCHT, C. M. *see* KASPER, J. S. *et al.* (1950). 117

MACGILLAVRY, C. H. and STRIJK, B. (1946*a*) *Nature* **157**, 135. 115
— and — (1946*b*) *Physica* **11**, 369. 115
— and — (1946*c*) *Physica* **12**, 129. 115

MARTIN, D. L. *see* NEWKIRK, J. B. *et al.* 114
MASON, W. P. (1950) *Piezoelectric Crystals and their Applications to Ultrasonics.* New York, van Nostrand. 181
MATSUBRA, T. (1952) *J. Phys. Soc. Japan* **7**, 270. 3
MAUGUIN, C. (1928) *C.R. Acad. Sci., Paris* **187**, 303. 113
— and LAVAL, J. (1939) *C.R. Acad. Sci., Paris* **208**, 1446. 58
MÉRING, J. (1949) *Acta cryst.* **2**, 371. 113
— *see* LONGUET-ESCARD, J. (1957). 114
MORANDI, G. *see* HOPPE, W. *et al.* (1957). 162, 163
MULDAWER, L. *see* SCHWARTZ, M. (1958). 59, 96
MURAKAMI, T. (1953*a*) *J. Phys. Soc. Japan* **8**, 31. 115
— (1953*b*) *J. Phys. Soc. Japan* **8**, 36. 115

NATH, N. S. N. (1943) *Nature* **151**, 196. 2
— *see* RAMAN, C. V. (1940*a*, *b*). 2
NEWKIRK, J. B., SMOLUCHOWSKI, R., GEISLER, A. H., and MARTIN, D. L. (1951) *Acta cryst.* **4**, 507. 114
NIEUWENKAMP, W. *see* BIJVOET, J. M. (1933). 113
NILAKANTAN, P. *see* RAMAN, C. V. (1940*a*, *b*, *c*; 1941*a*, *b*). 2
NYE, J. F. (1957) *Physical Properties of Crystals.* Oxford: Clarendon Press. 181

OGAWA, S. *see* HIRABAYASHI, M. (1956). 116
OLMER, P. (1948) Thesis. Paris. 2, 59, 96, 97, 98, 102, 103, 107
OUGHTON, B. M. *see* BRINDLEY, G. W. *et al.* (1950). 113
OWSTON, P. G. (1949) *Acta cryst.* **2**, 222. 173

PASHALOV, A. I. (1950) *Dokl. Akad. Nauk S.S.S.R.* **72**, 281. 114
PASKIN, A. (1958) *Acta cryst.* **11**, 165. 75
— *see* CHIPMAN, D. R. (1959*a*, *b*). 117
PATERSON, M. S. (1952) *J. appl. Phys.* **23**, 805. 3, 113
PEISER, H. S., ROOKSBY, H. P., and WILSON, A. J. C. (1955) *X-Ray Diffraction by Polycrystalline Materials.* The Institute of Physics. 55
PISHAROTY, P. R. (1941) *Proc. Indian Acad. Sci.* A **14**, 377. 2
— and SUBRAHMANIAN, R. V. (1941) *Proc. Indian Acad. Sci.* A **14**, 439. 2
POPE, N. K. (1949) *Acta cryst.* **2**, 325. 59
POWELL, H. M. and HUSE, G. (1943) *J. Chem. Soc.* **47**, 435. 117
PRASAD, S. C. and WOOSTER, W. A. (1955*a*) *Acta cryst.* **8**, 361. 59
— and — (1955*b*) *Acta cryst.* **8**, 506. 59
— and — (1955*c*) *Acta cryst.* **8**, 614. 90
— and — (1955*d*) *Acta cryst.* **8**, 682. 59
— and — (1956*a*) *Acta cryst.* **9**, 38. 59, 78, 83, 86, 116
— and — (1956*b*) *Acta cryst.* **9**, 169. 59, 76
— and — (1956*c*) *Acta cryst.* **9**, 304. 69
PRESTON, G. D. (1938*a*) *Proc. Roy. Soc.* A **167**, 526. 114, 140
— (1938*b*) *Phil. Mag.* **26**, 855. 114, 140
PRINCE, E. and WOOSTER, W. A. (1951) *Acta cryst.* **4**, 191. 59
— and — (1953) *Acta cryst.* **6**, 450. 59, 88, 90

VAN RAIJEN, L. L. (1944) *Physica* **11**, 114. 3
RAMACHANDRAN, G. N. (1949) Thesis. Cambridge. 174

RAMACHANDRAN, G. N. and WOOSTER, W. A. (1949) *Nature* **164**, 839. 59
— and — (1951*a*) *Acta cryst.* **4**, 335. 97, 99, 175
— and — (1951*b*) *Acta cryst.* **4**, 431. 59, 157, 158
— *see* KRISHNAN, R. S. (1945). 2
RAMAN, C. V. (1941*a*) *Proc. Indian Acad. Sci.* A **13**, 1. 2
— (1941*b*) *Proc. Indian Acad. Sci.* A **14**, 317. 2
— (1942) *Proc. Roy. Soc.* A **179**, 289. 2
— (1948) *Nature* **162**, 23. 2
— (1955) *Proc. Indian Acad. Sci.* A **42**, 163. 2, 56
— (1958*a*) *Proc. Indian Acad. Sci.* A **47**, 263. 2
— (1958*b*) *Proc. Indian Acad. Sci.* A **47**, 335. 2
— (1958*c*) *Proc. Indian Acad. Sci.* A **48**, 1. 2
— and NATH, N. S. N. (1940*a*) *Proc. Indian Acad. Sci.* A **12**, 83. 2
— and — (1940*b*) *Proc. Indian Acad. Sci.* A **12**, 427. 2
— and NILAKANTAN, P. (1940*a*) *Proc. Indian Acad. Sci.* A **11**, 379. 2
— and — (1940*b*) *Proc. Indian Acad. Sci.* A **12**, 141. 2
— and — (1940*c*) *Curr. Sci.* **9**, 165. 2
— and — (1941*a*) *Proc. Indian Acad. Sci.* A **14**, 317. 2
— and — (1941*b*) *Curr. Sci.* **10**, 241. 2
RIDLEY, P. *see* BRINDLEY, G. W. (1938). 1
ROBERTS, B. W. *see* WARREN, B. E. *et al.* (1951). 146, 150, 152
ROBINSON, K. *see* BRINDLEY, G. W. *et al.* (1950). 113
ROOKSBY, H. P. *see* PEISER, H. S. *et al.* (1955). 55
RUDMAN, P. S. *see* FLINN, P. A. *et al.* (1954). 116

SÁNDOR, E. E. (1960) Thesis. Cambridge University. 166, 173
— and WOOSTER, W. A. (1959) *Acta cryst.* **12**, 332. 59
SARGINSON, K. (1942) *Proc. Roy. Soc.* A **180**, 305. 59
— *see* BORN, M. (1941). 2
SAUNDER, D. H. *see* JAMES, R. W. (1947). 117
— *see* — (1948). 117
SCHAEFER, CL. and BERGMANN, L. (1934) *Sitz. Ber. berl. Akad. Wiss. Phys. Math. Kl.* **13**, 192. 57
— and — (1935 *Sitz. Ber. berl. Akad. Wiss. Phys. Math. Kl.* **14**, 222. 57
SCHWARTZ, M. and MULDAWER, L. (1958) *J. appl. Phys.* **29**, 1561. 59, 96
SEITZ, F. (1943) *The Physics of Metals.* New York: McGraw-Hill. 181
SHÔJI, H. (1933) *Z. Kryst.* **84**, 74. 113
SIEGEL, S. (1941) *Phys. Rev.* **59**, 371. 59
— and ZACHARIASEN, W. H. (1940) *Phys. Rev.* **57**, 795. 1, 59
SMITH, H. *see* LONSDALE, K. (1941). 59
SMOLUCHOWSKI, R. *see* NEWKIRK, J. B. *et al.* (1951). 114
SOKOLNIKOFF, I. S. (1946) *Mathematical Theory of Elasticity.* New York: McGraw-Hill. 181
STEEPLE, A. and EDMUNDS, I. G. (1956) *Acta cryst.* **9**, 934. 116
STRIJK, B. *see* MACGILLAVRY, C. H. (1946*a*, *b*, *c*). 115
SUBRAHMANIAN, R. V. *see* PISHAROTY, P. R. (1941). 2
SUITA, H. *see* TANAKA, K. *et al.* (1959). 54
SUONINEN, E. and WARREN, B. E. (1958) *Acta metallurgica.* **6**, 172. 117, 146
SUZUKI, K. (1958) *J. Phys. Soc. Japan* **13**, 179. 115
SYKES, C. and JONES, F. W. (1936) *Proc. Roy. Soc.* A **157**, 213. 115
— *see* JONES, F. W. (1938). 115

TANAKA, K., KATAYAMA, K., CHIKAWA, J., and SUITA, H. (1959) *Rev. sci. Instrum.* **30**, 430. 54
TAYLOR, C. A., HINDE, R. M., and LIPSON, H. (1951) *Acta cryst.* **4**, 261. 115, 138
— *see* LIPSON, H. (1951). 155
— *see* — (1958). 4
TAYLOR, W. J. (1951*a*) *Phys. Rev.* **82**, 279. 2
— (1951*b*) *Phys. Rev.* **84**, 148. 2
TELLER, E. *see* HENDRICKS, S. B. (1942). 113
TOMAN, K. (1955) *Acta cryst.* **8**, 578. 115, 140, 142
— (1957) *Acta cryst.* **10**, 187. 115, 140, 142

UMANSKII, M. M. (1960) *Apparatus for Crystal Structure Investigation.* Moscow: State Publications on Physicomathematical literature. 25, 55

VISWANATHAN, K. S. (1954) *Proc. Indian Acad. Sci.* A **39**, 196. 56
VOIGT, W. (1910) *Lehrbuch der Kristallphysik.* Leipzig: B. G. Teubner. 59, 181

WAGNER, C. N. J. (1957*a*) *Acta metallurgica* **5**, 427. 116
— (1957*b*) *Acta metallurgica* **5**, 477. 117
WALKER, C. B. (1952) *J. appl. Phys.* **23**, 118. 3, 115
— (1956) *Phys. Rev.* **103**, 547. 2, 55, 59, 96, 109
WALLER, I. (1923) *Z. Phys.* **17**, 398. 1
— (1925) Dissertation. Uppsala. 1
— (1928) *Z. Phys.* **51**, 213. 1
WARREN, B. E., AVERBACH, B. L., and ROBERTS, B. W. (1951) *J. appl. Phys.* **22**, 1493. 146, 150, 152
— *see* BORIE, B. (1956). 116
— *see* CHIPMAN, D. (1950). 54, 116
— *see* COLE, H. (1952). 2, 96, 112
— *see* SUONINEN, E. (1958). 117, 146
WASASTJERNA, J. A. (1947) *Soc. Sci. Fennica. Commentationes Phys. Math.* **13**, No. 5. 146
WILLIAMS, E. J. *see* BRAGG, W. L. (1935). 145
WILLIS, B. T. M. (1957*a*) *Proc. Roy. Soc.* A **239**, 184. 117
— (1957*b*) *Proc. Roy. Soc.* A **239**, 192. 117
— (1958) *Proc. Roy. Soc.* A **248**, 183. 118, 119, 120, 126, 127
— (1959) *Acta cryst.* **12**, 683. 114
WILMAN, H. *see* CHARLESBY, A. *et al.* (1939). 164
WILSON, A. J. C. (1942) *Proc. Roy. Soc.* A **180**, 277. 113
— (1947) *Nature* **160**, 304. 115
— (1949*a*) *X-ray Optics.* London: Methuen. 115, 126, 129, 132, 139
— (1949*b*) *Research* **2**, 541. 117
— (1952) *Acta cryst.* **5**, 318. 3, 117
— (1955) *Nuovo Cimento* **1**, 277. 117
— (1955) *see* PEISER, H. S. *et al.* 55
WOOSTER, W. A. (1938) *A Textbook on Crystal Physics.* Cambridge University Press. 85, 181
— (1956) *Proc. Nat. Acad. Sci. (India)* **25**, 58. 128
— (1960*a*) *Kristallographia* **5**, 375. 37
— (1960*b*) *Kristallographia* **5**, 788. 27
— *see* HOERNI, J. (1952*a*). 114

WOOSTER, W. A. *see* HOERNI, J. (1952*b*). 25, 59
—— *see* —— (1952*c*). 31, 177
—— *see* —— (1953). 37
—— *see* —— (1955). 114, 127, 132
—— *see* JOEL, N. (1960). 56
—— *see* PRASAD, S. C. (1955*a*). 59
—— *see* —— (1955*b*). 59
—— *see* —— (1955*c*). 90
—— *see* —— (1955*d*,. 59
—— *see* —— (1956*a*). 59, 78, 83, 86, 116
—— *see* —— (1956*b*). 59, 76
—— *see* —— (1956*c*). 69
—— *see* PRINCE, E. (1951). 59
—— *see* —— (1953). 59, 88, 90
—— *see* RAMACHANDRAN, G. N. (1949). 59
—— *see* —— (1951*a*). 97, 99, 175
—— *see* —— (1951*b*). 59, 157, 158
—— *see* SÁNDOR, E. (1959). 59

ZACHARIASEN, W. H. (1940) *Phys. Rev.* **57**, 597. 1, 59
—— (1945) *Theory of X-ray Diffraction in Crystals*. New York: Wiley & Sons. 59, 146, 147
—— (1947) *Phys. Rev.* **71**, 715. 113
—— *see* SIEGEL, S. (1940). 1, 59

INDEX OF SYMBOLS

$\bar{\psi}$	angle between the plane containing the axes $\bar{X}$ and $\bar{V}$ and the plane containing the axis $\bar{X}$ and the rekha PQ	178
ω	angular velocity of rotation of the crystal	166
ω_{ij}	librational components of the tensor defining atomic thermal movements in a molecule	156
Ω	the solid angle of the beam admitted by the window in front of the detector	54

SUBJECT INDEX

A CATALOG OF SELECTED

DOVER BOOKS

IN SCIENCE AND MATHEMATICS

RELATIVITY, THERMODYNAMICS AND COSMOLOGY, Richard C. Tolman. Landmark study extends thermodynamics to special, general relativity; also applications of relativistic mechanics, thermodynamics to cosmological models. 501pp. 5⅜ x 8½. 65383-8 Pa. $13.95

APPLIED ANALYSIS, Cornelius Lanczos. Classic work on analysis and design of finite processes for approximating solution of analytical problems. Algebraic equations, matrices, harmonic analysis, quadrature methods, much more. 559pp. 5⅜ x 8½. 65656-X Pa. $13.95

INTRODUCTION TO ANALYSIS, Maxwell Rosenlicht. Unusually clear, accessible coverage of set theory, real number system, metric spaces, continuous functions, Riemann integration, multiple integrals, more. Wide range of problems. Undergraduate level. Bibliography. 254pp. 5⅜ x 8½. 65038-3 Pa. $8.95

INTRODUCTION TO QUANTUM MECHANICS With Applications to Chemistry, Linus Pauling & E. Bright Wilson, Jr. Classic undergraduate text by Nobel Prize winner applies quantum mechanics to chemical and physical problems. Numerous tables and figures enhance the text. Chapter bibliographies. Appendices. Index. 468pp. 5⅜ x 8½. 64871-0 Pa. $12.95

ASYMPTOTIC EXPANSIONS OF INTEGRALS, Norman Bleistein & Richard A. Handelsman. Best introduction to important field with applications in a variety of scientific disciplines. New preface. Problems. Diagrams. Tables. Bibliography. Index. 448pp. 5⅜ x 8½. 65082-0 Pa. $12.95

MATHEMATICS APPLIED TO CONTINUUM MECHANICS, Lee A. Segel. Analyzes models of fluid flow and solid deformation. For upper-level math, science and engineering students. 608pp. 5⅜ x 8½. 65369-2 Pa. $14.95

ELEMENTS OF REAL ANALYSIS, David A. Sprecher. Classic text covers fundamental concepts, real number system, point sets, functions of a real variable, Fourier series, much more. Over 500 exercises. 352pp. 5⅜ x 8½. 65385-4 Pa. $11.95

PHYSICAL PRINCIPLES OF THE QUANTUM THEORY, Werner Heisenberg. Nobel Laureate discusses quantum theory, uncertainty, wave mechanics, work of Dirac, Schroedinger, Compton, Wilson, Einstein, etc. 184pp. 5⅜ x 8½. 60113-7 Pa. $6.95

INTRODUCTORY REAL ANALYSIS, A.N. Kolmogorov, S.V. Fomin. Translated by Richard A. Silverman. Self-contained, evenly paced introduction to real and functional analysis. Some 350 problems. 403pp. 5⅜ x 8½. 61226-0 Pa. $10.95

PROBLEMS AND SOLUTIONS IN QUANTUM CHEMISTRY AND PHYSICS, Charles S. Johnson, Jr. and Lee G. Pedersen. Unusually varied problems, detailed solutions in coverage of quantum mechanics, wave mechanics, angular momentum, molecular spectroscopy, scattering theory, more. 280 problems plus 139 supplementary exercises. 430pp. 6½ x 9¼. 65236-X Pa. $13.95

THE ELECTROMAGNETIC FIELD, Albert Shadowitz. Comprehensive undergraduate text covers basics of electric and magnetic fields, builds up to electromagnetic theory. Also related topics, including relativity. Over 900 problems. 768pp. 5⅝ x 8¼. 65660-8 Pa. $18.95

FOURIER SERIES, Georgi P. Tolstov. Translated by Richard A. Silverman. A valuable addition to the literature on the subject, moving clearly from subject to subject and theorem to theorem. 107 problems, answers. 336pp. 5⅜ x 8½. 63317-9 Pa. $9.95

THEORY OF ELECTROMAGNETIC WAVE PROPAGATION, Charles Herach Papas. Graduate-level study discusses the Maxwell field equations, radiation from wire antennas, the Doppler effect and more. xiii + 244pp. 5⅜ x 8½. 65678-0 Pa. $6.95

DISTRIBUTION THEORY AND TRANSFORM ANALYSIS: An Introduction to Generalized Functions, with Applications, A.H. Zemanian. Provides basics of distribution theory, describes generalized Fourier and Laplace transformations. Numerous problems. 384pp. 5⅜ x 8½. 65479-6 Pa. $11.95

THE PHYSICS OF WAVES, William C. Elmore and Mark A. Heald. Unique overview of classical wave theory. Acoustics, optics, electromagnetic radiation, more. Ideal as classroom text or for self-study. Problems. 477pp. 5⅜ x 8½. 64926-1 Pa. $13.95

CALCULUS OF VARIATIONS WITH APPLICATIONS, George M. Ewing. Applications-oriented introduction to variational theory develops insight and promotes understanding of specialized books, research papers. Suitable for advanced undergraduate/graduate students as primary, supplementary text. 352pp. 5⅜ x 8½. 64856-7 Pa. $9.95

A TREATISE ON ELECTRICITY AND MAGNETISM, James Clerk Maxwell. Important foundation work of modern physics. Brings to final form Maxwell's theory of electromagnetism and rigorously derives his general equations of field theory. 1,084pp. 5⅜ x 8½. 60636-8, 60637-6 Pa., Two-vol. set $25.90

AN INTRODUCTION TO THE CALCULUS OF VARIATIONS, Charles Fox. Graduate-level text covers variations of an integral, isoperimetrical problems, least action, special relativity, approximations, more. References. 279pp. 5⅜ x 8½. 65499-0 Pa. $8.95

HYDRODYNAMIC AND HYDROMAGNETIC STABILITY, S. Chandrasekhar. Lucid examination of the Rayleigh-Benard problem; clear coverage of the theory of instabilities causing convection. 704pp. 5⅜ x 8¼. 64071-X Pa. $14.95

CALCULUS OF VARIATIONS, Robert Weinstock. Basic introduction covering isoperimetric problems, theory of elasticity, quantum mechanics, electrostatics, etc. Exercises throughout. 326pp. 5⅜ x 8½. 63069-2 Pa. $9.95

DYNAMICS OF FLUIDS IN POROUS MEDIA, Jacob Bear. For advanced students of ground water hydrology, soil mechanics and physics, drainage and irrigation engineering and more. 335 illustrations. Exercises, with answers. 784pp. 6⅛ x 9¼. 65675-6 Pa. $19.95

ORDINARY DIFFERENTIAL EQUATIONS, Morris Tenenbaum and Harry Pollard. Exhaustive survey of ordinary differential equations for undergraduates in mathematics, engineering, science. Thorough analysis of theorems. Diagrams. Bibliography. Index. 818pp. 5⅜ x 8½. 64940-7 Pa. $18.95

STATISTICAL MECHANICS: Principles and Applications, Terrell L. Hill. Standard text covers fundamentals of statistical mechanics, applications to fluctuation theory, imperfect gases, distribution functions, more. 448pp. 5⅜ x 8½. 65390-0 Pa. $11.95

ORDINARY DIFFERENTIAL EQUATIONS AND STABILITY THEORY: An Introduction, David A. Sánchez. Brief, modern treatment. Linear equation, stability theory for autonomous and nonautonomous systems, etc. 164pp. 5⅝ x 8¼. 63828-6 Pa. $6.95

THIRTY YEARS THAT SHOOK PHYSICS: The Story of Quantum Theory, George Gamow. Lucid, accessible introduction to influential theory of energy and matter. Careful explanations of Dirac's anti-particles, Bohr's model of the atom, much more. 12 plates. Numerous drawings. 240pp. 5⅜ x 8½. 24895-X Pa. $7.95

THEORY OF MATRICES, Sam Perlis. Outstanding text covering rank, nonsingularity and inverses in connection with the development of canonical matrices under the relation of equivalence, and without the intervention of determinants. Includes exercises. 237pp. 5⅜ x 8½. 66810-X Pa. $8.95

GREAT EXPERIMENTS IN PHYSICS: Firsthand Accounts from Galileo to Einstein, edited by Morris H. Shamos. 25 crucial discoveries: Newton's laws of motion, Chadwick's study of the neutron, Hertz on electromagnetic waves, more. Original accounts clearly annotated. 370pp. 5⅜ x 8½. 25346-5 Pa. $10.95

INTRODUCTION TO PARTIAL DIFFERENTIAL EQUATIONS WITH APPLICATIONS, E.C. Zachmanoglou and Dale W. Thoe. Essentials of partial differential equations applied to common problems in engineering and the physical sciences. Problems and answers. 416pp. 5⅜ x 8½. 65251-3 Pa. $11.95

BURNHAM'S CELESTIAL HANDBOOK, Robert Burnham, Jr. Thorough guide to the stars beyond our solar system. Exhaustive treatment. Alphabetical by constellation: Andromeda to Cetus in Vol. 1; Chamaeleon to Orion in Vol. 2; and Pavo to Vulpecula in Vol. 3. Hundreds of illustrations. Index in Vol. 3. 2,000pp. 6⅛ x 9¼. 23567-X, 23568-8, 23673-0 Pa., Three-vol. set $44.85

CHEMICAL MAGIC, Leonard A. Ford. Second Edition, Revised by E. Winston Grundmeier. Over 100 unusual stunts demonstrating cold fire, dust explosions, much more. Text explains scientific principles and stresses safety precautions. 128pp. 5⅜ x 8½. 67628-5 Pa. $5.95

AMATEUR ASTRONOMER'S HANDBOOK, J.B. Sidgwick. Timeless, comprehensive coverage of telescopes, mirrors, lenses, mountings, telescope drives, micrometers, spectroscopes, more. 189 illustrations. 576pp. 5⅝ x 8¼. (Available in U.S. only) 24034-7 Pa. $11.95

SPECIAL FUNCTIONS, N.N. Lebedev. Translated by Richard Silverman. Famous Russian work treating more important special functions, with applications to specific problems of physics and engineering. 38 figures. 308pp. 5⅜ x 8½. 60624-4 Pa. $9.95

OBSERVATIONAL ASTRONOMY FOR AMATEURS, J.B. Sidgwick. Mine of useful data for observation of sun, moon, planets, asteroids, aurorae, meteors, comets, variables, binaries, etc. 39 illustrations. 384pp. 5⅜ x 8¼. (Available in U.S. only) 24033-9 Pa. $8.95

INTEGRAL EQUATIONS, F.G. Tricomi. Authoritative, well-written treatment of extremely useful mathematical tool with wide applications. Volterra Equations, Fredholm Equations, much more. Advanced undergraduate to graduate level. Exercises. Bibliography. 238pp. 5⅜ x 8½. 64828-1 Pa. $8.95

POPULAR LECTURES ON MATHEMATICAL LOGIC, Hao Wang. Noted logician's lucid treatment of historical developments, set theory, model theory, recursion theory and constructivism, proof theory, more. 3 appendixes. Bibliography. 1981 edition. ix + 283pp. 5⅜ x 8½. 67632-3 Pa. $8.95

MODERN NONLINEAR EQUATIONS, Thomas L. Saaty. Emphasizes practical solution of problems; covers seven types of equations. ". . . a welcome contribution to the existing literature...."–*Math Reviews*. 490pp. 5⅜ x 8½. 64232-1 Pa. $13.95

FUNDAMENTALS OF ASTRODYNAMICS, Roger Bate et al. Modern approach developed by U.S. Air Force Academy. Designed as a first course. Problems, exercises. Numerous illustrations. 455pp. 5⅜ x 8½. 60061-0 Pa. $10.95

INTRODUCTION TO LINEAR ALGEBRA AND DIFFERENTIAL EQUATIONS, John W. Dettman. Excellent text covers complex numbers, determinants, orthonormal bases, Laplace transforms, much more. Exercises with solutions. Undergraduate level. 416pp. 5⅜ x 8½. 65191-6 Pa. $11.95

INCOMPRESSIBLE AERODYNAMICS, edited by Bryan Thwaites. Covers theoretical and experimental treatment of the uniform flow of air and viscous fluids past two-dimensional aerofoils and three-dimensional wings; many other topics. 654pp. 5⅜ x 8½. 65465-6 Pa. $16.95

INTRODUCTION TO DIFFERENCE EQUATIONS, Samuel Goldberg. Exceptionally clear exposition of important discipline with applications to sociology, psychology, economics. Many illustrative examples; over 250 problems. 260pp. 5⅜ x 8½. 65084-7 Pa. $8.95

LAMINAR BOUNDARY LAYERS, edited by L. Rosenhead. Engineering classic covers steady boundary layers in two- and three- dimensional flow, unsteady boundary layers, stability, observational techniques, much more. 708pp. 5⅜ x 8½. 65646-2 Pa. $18 95

LECTURES ON CLASSICAL DIFFERENTIAL GEOMETRY, Second Edition, Dirk J. Struik. Excellent brief introduction covers curves, theory of surfaces, fundamental equations, geometry on a surface, conformal mapping, other topics. Problems. 240pp. 5⅜ x 8½. 65609-8 Pa. $8.95

GEOMETRY OF COMPLEX NUMBERS, Hans Schwerdtfeger. Illuminating, widely praised book on analytic geometry of circles, the Moebius transformation, and two-dimensional non-Euclidean geometries. 200pp. 5⅝ x 8¼. 63830-8 Pa. $8.95

MECHANICS, J.P. Den Hartog. A classic introductory text or refresher. Hundreds of applications and design problems illuminate fundamentals of trusses, loaded beams and cables, etc. 334 answered problems. 462pp. 5⅜ x 8½. 60754-2 Pa. $11.95

TOPOLOGY, John G. Hocking and Gail S. Young. Superb one-year course in classical topology. Topological spaces and functions, point-set topology, much more. Examples and problems. Bibliography. Index. 384pp. 5⅝ x 8¼. 65676-4 Pa. $10.95

STRENGTH OF MATERIALS, J.P. Den Hartog. Full, clear treatment of basic material (tension, torsion, bending, etc.) plus advanced material on engineering methods, applications. 350 answered problems. 323pp. 5⅜ x 8½. 60755-0 Pa. $9.95

ELEMENTARY CONCEPTS OF TOPOLOGY, Paul Alexandroff. Elegant, intuitive approach to topology from set-theoretic topology to Betti groups; how concepts of topology are useful in math and physics. 25 figures. 57pp. 5⅜ x 8½. 60747-X Pa. $3.95

ADVANCED STRENGTH OF MATERIALS, J.P. Den Hartog. Superbly written advanced text covers torsion, rotating disks, membrane stresses in shells, much more. Many problems and answers. 388pp. 5⅜ x 8½. 65407-9 Pa. $10.95

COMPUTABILITY AND UNSOLVABILITY, Martin Davis. Classic graduate-level introduction to theory of computability, usually referred to as theory of recurrent functions. New preface and appendix. 288pp. 5⅜ x 8½. 61471-9 Pa. $8.95

GENERAL CHEMISTRY, Linus Pauling. Revised 3rd edition of classic first-year text by Nobel laureate. Atomic and molecular structure, quantum mechanics, statistical mechanics, thermodynamics correlated with descriptive chemistry. Problems. 992pp. 5⅜ x 8½. 65622-5 Pa. $19.95

AN INTRODUCTION TO MATRICES, SETS AND GROUPS FOR SCIENCE STUDENTS, G. Stephenson. Concise, readable text introduces sets, groups, and most importantly, matrices to undergraduate students of physics, chemistry, and engineering. Problems. 164pp. 5⅜ x 8½. 65077-4 Pa. $7.95

THE HISTORICAL BACKGROUND OF CHEMISTRY, Henry M. Leicester. Evolution of ideas, not individual biography. Concentrates on formulation of a coherent set of chemical laws. 260pp. 5⅜ x 8½. 61053-5 Pa. $8.95

THE PHILOSOPHY OF MATHEMATICS: An Introductory Essay, Stephan Körner. Surveys the views of Plato, Aristotle, Leibniz & Kant concerning propositions and theories of applied and pure mathematics. Introduction. Two appendices. Index. 198pp. 5⅜ x 8½. 25048-2 Pa. $8.95

THE DEVELOPMENT OF MODERN CHEMISTRY, Aaron J. Ihde. Authoritative history of chemistry from ancient Greek theory to 20th-century innovation. Covers major chemists and their discoveries. 209 illustrations. 14 tables. Bibliographies. Indices. Appendices. 851pp. 5⅜ x 8½. 64235-6 Pa. $18.95

CHALLENGING MATHEMATICAL PROBLEMS WITH ELEMENTARY SOLUTIONS, A.M. Yaglom and I.M. Yaglom. Over 170 challenging problems on probability theory, combinatorial analysis, points and lines, topology, convex polygons, many other topics. Solutions. Total of 445pp. 5⅜ x 8½. Two-vol. set.
Vol. I: 65536-9 Pa. $7.95
Vol. II: 65537-7 Pa. $7.95

FIFTY CHALLENGING PROBLEMS IN PROBABILITY WITH SOLUTIONS, Frederick Mosteller. Remarkable puzzlers, graded in difficulty, illustrate elementary and advanced aspects of probability. Detailed solutions. 88pp. 5⅜ x 8½.
65355-2 Pa. $4.95

EXPERIMENTS IN TOPOLOGY, Stephen Barr. Classic, lively explanation of one of the byways of mathematics. Klein bottles, Moebius strips, projective planes, map coloring, problem of the Koenigsberg bridges, much more, described with clarity and wit. 43 figures. 210pp. 5⅜ x 8½. 25933-1 Pa. $6.95

RELATIVITY IN ILLUSTRATIONS, Jacob T. Schwartz. Clear nontechnical treatment makes relativity more accessible than ever before. Over 60 drawings illustrate concepts more clearly than text alone. Only high school geometry needed. Bibliography. 128pp. 6⅛ x 9¼. 25965-X Pa. $7.95

AN INTRODUCTION TO ORDINARY DIFFERENTIAL EQUATIONS, Earl A. Coddington. A thorough and systematic first course in elementary differential equations for undergraduates in mathematics and science, with many exercises and problems (with answers). Index. 304pp. 5⅜ x 8½. 65942-9 Pa. $8.95

FOURIER SERIES AND ORTHOGONAL FUNCTIONS, Harry F. Davis. An incisive text combining theory and practical example to introduce Fourier series, orthogonal functions and applications of the Fourier method to boundary-value problems. 570 exercises. Answers and notes. 416pp. 5⅜ x 8½. 65973-9 Pa. $11.95

AN INTRODUCTION TO ALGEBRAIC STRUCTURES, Joseph Landin. Superb self-contained text covers "abstract algebra": sets and numbers, theory of groups, theory of rings, much more. Numerous well-chosen examples, exercises. 247pp. 5⅜ x 8½.
65940-2 Pa. $8.95

STARS AND RELATIVITY, Ya. B. Zel'dovich and I. D. Novikov. Vol. 1 of *Relativistic Astrophysics* by famed Russian scientists. General relativity, properties of matter under astrophysical conditions, stars and stellar systems. Deep physical insights, clear presentation. 1971 edition. References. 544pp. 5⅜ x 8½.
69424-0 Pa. $14.95

Prices subject to change without notice.

Available at your book dealer or write for free Mathematics and Science Catalog to Dept. GI, Dover Publications, Inc., 31 East 2nd St., Mineola, N.Y. 11501. Dover publishes more than 250 books each year on science, elementary and advanced mathematics, biology, music, art, literature, history, social sciences and other areas.